Innovation, Strategy and Risk in Construction

Turning Serendipity into Capability

Innovation, Strategy and Risk in Construction integrates insights from business and government leaders with contemporary research, to help built environment professionals turn serendipity to their own advantage by building greater innovative and adaptive capacity into their operations. Accessible and full of practical examples, the book argues that traditional business strategies which seek to systematise innovation and eliminate uncertainty need to be balanced with more flexible approaches which acknowledge and harness uncertainty.

The missing key to innovation, it is argued, is to turn serendipity into capability. The author proposes a simple model which allows managers to tap into the increasingly dynamic and interconnected nature of the construction industry. Innovation does not occur in isolation within individual firms, but through collaboration. Each stakeholder in the construction industry has a responsibility to drive innovation, and this book will be key reading for consultants, contractors, subcontractors, suppliers and clients, as well as policy makers and all serious students of construction management.

Martin Loosemore is a Professor at UNSW, Sydney, Australia and a Visiting Professor at Loughborough University, UK. Martin was an advisor on workplace productivity and reform to the Cole Royal Commission into the Building and Construction Industry and a member of Australian Government's Built Environment Industry Innovation Council.

Innovation, Strategy and Risk in Construction

Turning Serendipity into Capability

Martin Loosemore

LONDON AND NEW YORK

First published 2014
by Routledge
2 Park Square, Milton Park, Abingdon, Oxon, OX14 4RN

and by Routledge
711 Third Avenue, New York, NY 10017

Routledge is an imprint of the Taylor & Francis Group, an informa business

British Library Cataloguing in Publication Data
A catalogue record for this book is available from the British Library

Library of Congress Cataloging-in-Publication Data
A catalog record has been requested for this book

ISBN13: 978-0-415-67600-7 (hbk)
ISBN13: 978-0-415-67599-4 (pbk)
ISBN13: 978-0-203-80915-0 (ebk)

Typeset in Bembo by
Keystroke, Station Road, Codsall, Wolverhampton

Printed and bound in Great Britain by
TJ International Ltd, Padstow, Cornwall

To Tristan, Bryony and Elliot.

Contents

List of figures

List of tables

Acknowledgements

I would like to thank the following innovators and entrepreneurs who were acknowledged by peers as leaders in their field and who kindly gave their time to 'reality-check' the ideas in this book.

Michael Argyrou, Joint Managing Director, Co-founder, Hickory Group
Don Aroney, Executive Director Operations, Brookfield Multiplex
Peter Barda, Executive Director, Australian Construction Industry Forum
Doron Ben-Meir, CEO, Commercialisation Australia, Australian Government
Tricia Berman, General Manager, Innovation Policy Branch, Department of Industry, Innovation, Science, Research and Tertiary Education, Australian Government
Felicity Brazil, Continuous Improvement Network, Enterprise Connect, Department of Industry, Innovation, Science, Research and Tertiary Education, Australian Government
Dr Venny Chandra, Sustainable Development Advisor and Entrepreneur, Indonesia.
Dr (HC) Ir. Ciputra, Founder Ciputra Group & UC (Universitas Ciputra), Indonesia
Andrew Council, Group Head of Construction Risk, Lend Lease.
Glenn Corke, Managing Director, Donald Cant Watts Corke
Dr Tom Crow, Managing Director, Crowbar Pty Ltd
Chris Cusack, Director, Donald Cant Watts Corke
Andrew Ferguson, Coverforce Pty Ltd, Past State Secretary, Construction, Forestry, Mining and Energy Union (CFMEU), NSW Branch, Construction & General Division
John Flecker, CEO, Brookfield Multiplex Australasia
Laurie Foy, Regional Director, Construction and Development, Brookfield Multiplex
Dr James Gardiner, Lead of Design & Innovation, The Engineering Excellence Group, Laing O'Rourke
David Ghannoum, Regional Managing Director NSW, Construction and Development, Brookfield Multiplex
Dr Michael Green, General Manager Manufacturing Innovation, Department

of Industry, Innovation, Science, Research and Tertiary Education, Australian Government

Dave Higgon, Employee Relations Manager, Brookfield Multiplex

Justin Hill, Manager, Venture Capital Policy Section, Innovation Policy Branch, Department of Industry, Innovation, Science, Research and Tertiary Education, Australian Government

Gary Horwitz, Head of Retail – Development and Asset Management, Lend Lease

Richard Hough, Principal, Arup, Sydney

Cathy Inglis, Group Technical, Research and Engineering Manager, Brickworks Building Products

Brian Kooyman, Chairman of the Board and CEO of Confluence Tracey Brunstrom & Hammond (TBH), Director for International Operations TBH

George Kostas, Managing Director, Construction and Development, Brookfield Multiplex

Matthew McCann, Managing Partner, Brookfield Financial

John V. McCarthy AO, Chairman, AEH Property Group

Dr Andrew Maher, Leader, Digital Innovation, Arup

Geoff Moore, Head of Commercial, Construction Centre of Excellence, Lend Lease

Dave Noonan, National Secretary of the Construction, Forestry, Mining and Energy Union (CFMEU), Construction and General Division

Darren O'Connell, Strategic Marketing Manager, OneSteel

Charlie Reilly, Executive General Manager, Australia and New Zealand, DTZ; Vice Chairman, International Project Finance Association Australia

Matthew Squire, Manager, Policy Section, Business Improvement Branch, Enterprise Connect, Department of Industry, Innovation, Science, Research and Tertiary Education, Australian Government

Ray Thompson, General Manager, Innovation & Market Development, CSR Building Products

Jane Urquhart, General Manager, Business Improvement, Enterprise Connect, Department of Industry, Innovation, Science, Research and Tertiary Education, Australian Government

I would also like to acknowledge Sue Holliday, the Secretariat and co-members of the Australian Government's Built Environment Industry Innovation Council (BEIIC) whose enthusiasm and commitment to innovation inspired me to write this book.

1 Serendipity and innovation

> Once we rid ourselves of traditional thinking we can get on with creating the future.
>
> James Bertrand

Introduction – grass roots innovation

In its simple terms innovation is the process of bringing new creative ideas to reality. As Theodore Levitt, former Editor of the *Harvard Business Review* succinctly put it, creativity involves thinking up new things while innovation involves doing new things. A more formal definition of innovation is provided by the Organisation for Economic Co-operation and Development (OECD) which describes innovation as the creation of new products, services or business processes that create wealth or social welfare. This definition is much more useful for a number of reasons. First, it links innovation to 'value-creation'. Second, it recognizes that the concept of 'value' is much broader than the short-term economic interests of corporations. Third, it is a departure from traditional notions of innovation which portray it as a highly linear, laboratory-based process which is best measured by Research and Development (R&D) spending and the types of initiatives found in high-technology industries such as manufacturing. Innovation in the construction sector rarely happens in this way, meaning that the construction sector has been widely overlooked in national innovation statistics. This has left many with the misleading and highly damaging impression that it is a rather backward industry which is slow to change and adopt new ideas (Gambatese and Hallowell 2011).

As we will see in this book the construction sector does innovate. Furthermore, traditional ideas about innovation are outdated and many business leaders are becoming disillusioned with the low return on investment from their R&D activities. For example, while significant growth in the number of R&D personnel working in US firms over recent decades would appear to indicate a more innovative economy, the average R&D worker in 1950 contributed about seven times more to US total factor productivity than an equivalent worker did in 2000 (*The Economist* 2013). In response, many business leaders are adopting a more 'grass-roots' approach to innovation. This approach seeks to harness the talents and experience of their workforce and to capitalize on their unique relationships with key business partners who may have previously been seen as 'outsiders'. This is the approach to innovation which is advocated and explored in this book. Not only does it more closely reflect the way that innovation occurs in the construction sector, but it offers considerable potential in tackling the many challenges which face contemporary business and society.

A challenging future

Much of the recent interest in innovation has been driven by a growing chorus of economists and business leaders who believe that the world is running out of ideas to deal with the huge challenges which lie ahead. Recent 'Futures' reports like the World Economic Forum's *Global Risks 2013* are enough to make any manager crawl under a stone. Problems of chronic labour market imbalances, rampant climate change, depleting natural resources, contagious financial markets, computer and human viruses and digital wildfires which could reap global havoc are just some of the threats business and society faces in the near future (WEF 2013). According to *The Economist* (2013), the rate of global innovation in responding to such challenges has been declining for decades. Furthermore, the impact of future innovations are likely to pale in comparison to those of the past, such as electricity, internal combustion, plumbing, petrochemicals, the telephone and computers. According to Peter Theil, a founder of Paypal, innovation in the US is somewhere between dire straits and dead. And some commentators believe that the global financial crisis masks a deeper and much more disturbing 'stagnation' of new ideas in the business world. Such arguments are supported by productivity data over the last two centuries which show output per worker steadily growing at a rate of 1 per cent per year in Britain, Europe and America during the nineteenth century. By the middle of the twentieth century it was 2.5 per cent per year and average incomes doubled every generation. However, by the 1970s output per worker had dropped to 2 per cent per year and by 2000 it was back to 1 per cent. While optimists argue that rates of innovation have always ebbed and flowed and that the rise of the developing world will magnify the impact of future innovations by spreading the benefits across a larger number of people, pessimists argue that recent rates of progress in travel, living standards, communications and health have slowed considerably compared to the early and mid-twentieth century. Potential causes which have been muted include an insidious tide of poorly crafted government regulation which discourages risk taking and reductions in government funding for R&D. Globalization has also been blamed for encouraging firms to offshore their operations in the search for cheap labour rather than investing in productivity-improving innovations at home.

The innovation imperative

Whether one is an optimist or pessimist, contemporary business rhetoric is that managers 'must' innovate to survive. Business leaders are constantly told that innovation is 'essential' to business growth, competitiveness and productivity in a world of increasing competition and depleting natural resources. For example, a recent McKell Institute Report into productivity in Australian industry declares that firms must 'innovate or perish' (Green et al. 2012: 12). According to the report, executives are faced with a fundamental choice. They can either take 'the "low-road" of narrow cost-cutting and an unwinnable "race to the bottom", or the "high road" of longer term dynamic efficiency gains in a knowledge-based high-wage, high-productivity economy'. Evidence is presented to show that innovative businesses are twice as likely to report increased productivity compared with businesses that don't innovate. And this reaffirms other research by influential organizations such as the Grattan Institute which have identified a lack of innovation as a key area of reform needed to raise national productivity performance (Eslake and Walsh 2011).

In highly homogeneous industries like construction, where there is little to differentiate competitors, innovation is especially important. Many competitors in the construction sector draw from the same pool of suppliers, subcontractors and consultants. In this type of business environment, where cost-bases are broadly the same, innovation can provide significant competitive advantage by enabling firms to increase relative productivity or offer clients a better value-proposition to their clients. Yet, according to government statistics and many independent research projects, innovation appears to be concentrated in a small number of firms in high-tech industries such as manufacturing, electronics and pharmaceuticals, which are renowned as 'persistent' innovators. With very few exceptions, firms which operate within the construction industry are conspicuous by their absence from published lists of these firms. For example, BRW's recent list of the thirty most innovative businesses firms in Australia includes none from the construction sector (Mills 2012).

Innovation in the construction sector

Given the above, one could be forgiven for thinking that the construction sector is a low innovation sector. However, as we will see in this book, more recent research has begun to question this simplistic perspective, revealing that a considerable amount of innovation *does* occur in the construction industry, even though it is often hidden from view (NESTA 2007; Abbot et al. 2007). Although the construction industry undoubtedly has a 'long tail' of poorly managed firms where performance could be improved significantly, leading firms in the construction industry innovate daily to successfully deliver highly complex projects in dynamic and uncertain environments – a capacity which is justly admired by leading firms in many other industries. The construction industry clearly has a lot to be proud of and it is important to recognize this.

Furthermore, although construction is widely regarded as a hard-nosed and highly commercial industry, recent research shows that the drivers of innovation in the sector are not purely selfish. Increasingly, as Loosemore and Phua's (2011) critical analysis of corporate social responsibility in the construction sector showed, the construction industry is playing a critical role in addressing the world's most profound social and environmental challenges such as poverty, urbanization, pollution, resource depletion, population growth and climate change.

In this book we will discuss the many organizational, institutional, cultural and psychological challenges which leaders in the construction sector face in responding to these complex challenges. We will also discuss the political and social aspects of innovation in what Beck (1992) insightfully described as an increasingly paranoid 'risk society'. Today, the public is better informed, educated and empowered than at any time in history to understand and influence the impact of business activities on their lives. Consequently, the activities of many high-impact industries like construction have come under increasing public scrutiny, presenting new challenges for managers which are not well understood. It is clear that managers can no longer be absolved from the wider social and ecological implications of their decisions. The challenge then, is not only to increase levels of innovation in the construction sector, but to ensure that firms in the industry contribute to social well-being and are seen to be doing so. As Fox and Hooper (2011) correctly point out, today's true leaders operate at the intersection of business and society.

Evidence-based strategy

Like all fashionable topics in business, the subject of innovation has attracted its fair share of self-proclaimed gurus who have generated literally thousands of articles, books, websites and blogs. While there is certainly plenty of information out there, the problem for leaders is that much of it is based on nothing more than shallow opinion-based surveys, anecdote and hearsay. Many management consultants make a very good living out of this questionable information and it is no coincidence that the most dubious information often appears first in any web-based search on the subject of innovation. While personal experience and intuition is often interesting and might occasionally be of substance, it is very hard for managers to separate rhetoric from reality. Consequently, as a recent article by Lafley et al. (2012: 57) in the *Harvard Business Review* pointed out, 'few managers succeed in marrying empirical rigour with creative thinking'.

There is plenty of evidence to show that firms in the construction sector are equally susceptible to this deception and that many have mistakenly fallen for the miraculous promises of business success which accompany it. As Dainty and Loosemore's (2012) critical review of human resource management practices in the construction sector illustrates, the consequences of following this faulty advice can be catastrophic for the firms and employees affected. To avoid this happening, the aim of this book is to help managers working in the construction sector to sift the good information from the bad, so that they can base their innovation strategies, as far as possible, on fact rather than fiction.

Building agility

With so much being written about the subject of innovation, it is very easy to simply regurgitate what has already been written about the subject. What quickly becomes apparent when one reads the literature in this area is that it is highly repetitive. The normal approach is to portray innovation as a highly mechanistic and predictable process which is best planned well in advance. Uncertainty and unpredictability in the process are widely portrayed as a waste of resources and as a sign of managerial failure. Innovation is seen, like many other contemporary areas of management, as a key mechanism to help managers deliver ever more challenging growth targets and to satisfy the seemingly insatiable appetite of increasingly dispersed shareholders for short-term financial returns.

On the contrary, this book argues that innovation has a neglected political and social dimension and is not just about financial and productivity returns. Furthermore, in an increasingly dynamic, uncertain and interconnected business world, managers need to better embrace uncertainty and learn to recognize and harness the spontaneous and unpredictable nature of the innovation process. The alternative approach to innovation being offered in this book does not advocate the model of the 'mad professor' accidentally stumbling upon something new. And it doesn't argue that innovation should be left to chance. These approaches to innovation do not represent a practical and sustainable business model for any organization. However, this book does argue that planned approaches which seek to 'systemize' innovation need to be balanced with flexible strategies which can harness the largely untapped unpredictable and serendipitous opportunities offered by the new world order (or disorder). Serendipity means 'finding something when you are looking for something else' and as we will show in this book, many entrepreneurs say that serendipity has played a large part in their success. They openly admit that their first business plans were not always where they ended up. It wasn't their plans that led them to innovate. Rather, it was their open-mindedness, sensitivity, alertness and willingness to adapt those plans and respond to the unexpected.

This book is about how we can build this agility into the construction industry. It is of relevance to every firm that is involved in the industry from clients through consultants to contractors, subcontractors and suppliers. Based in rigorous research and in conversations with some of the industry's leading global innovators from across the supply chain, one of the resounding messages to emerge from this book is that innovation does not occur in isolation within individual firms but through collaboration. Everyone in the industry has a collective responsibility to drive innovation and a high dependency on others in the industry to do so. The idea of organizing for serendipity might seem like an oxymoron. And a business strategy which relies, even in part, on serendipity might seem like an unsustainable and somewhat frightening idea. This book will show that these responses are a consequence of an outdated way of thinking about business strategy which has developed to eliminate and suppress uncertainty, avoid risk raking and promote business unpredictability as a sign of managerial failure. The dangers of this approach are clearly evident in the construction sector. Nowhere is this more vividly illustrated than in the all-pervasive subcontracting model which has fragmented the construction industry, leading to a multitude of problems which include abuses of human rights, corruption, underinvestments in people and knowledge development and a confrontational culture of risk transfer where there is little incentive to innovate and where risk is passed to the point of least resistance and lowest capability (Dainty and Loosemore 2012). The negative effects of this traditional approach to management which is designed to suppress uncertainty have also been well documented by leading scholars outside the construction sector. For example, the well-known Lebanese American scholar Nassim Nicholas Taleb shows that the critical issue in the lead-up to the financial crisis was business and governments' artificial suppression of volatility (the ups-and-downs of life) in the name of economic and political stability. According to Taleb and Blyth (2011), what happened in the global financial crisis, and subsequently in the political upheavals of Libya, Egypt and Syria, is the inevitable result of suppressing risk in complex systems. It seems that complex systems, which have artificially suppressed volatility, become increasingly fragile over time while exhibiting no outward signs of stress until the hidden risks accumulate and explode into an uncontrollable and seemingly unpredictable 'Black Swan' event. According to Taleb and Blyth, managers and policy-makers who seek to reduce variability in complex systems unwittingly do the opposite by ironically increasing the chance that the system will eventually collapse.

This book builds on this line of thought by exploring how managers can better harness the opportunities which lie untapped in the problems of randomness, probability and uncertainty which increasingly characterize their decisions. In doing so it exposes the problems of rigid organizations that suppress unpredictability and illustrates the benefits of agile organizations which acknowledge the inevitability of unpredictability. By integrating rigorous research with the real-life insights of leading innovators in industry, it presents a new evidence-based model of innovation which can be used to drive the future development of industries like construction which are distinguished by their unpredictability.

Serendipity in the innovation process

Unfortunately, because of serendipity's association with the rather mystical, mysterious and unmeasurable concept of luck, the traditions of science and management have driven underground discussions about its role in the innovation process. Yet its contribution to the advancement of civilization throughout history is plain to see. Famous examples of where scientists or innovators seem to have stumbled on a new insight or idea are all too easy to find. These include:

1. Alexander Fleming discovered penicillin when a petri dish of bacteria accidentally became infected by a mould which killed the bacteria.
2. Vulcanized rubber was discovered when Charles Goodyear accidentally left a piece of rubber mixture with sulphur on a hot plate.
3. Safety glass was discovered when a French scientist accidentally knocked a glass flask to the floor and observed that the broken pieces were held together by a liquid plastic that had evaporated and formed a thin film inside the flask.
4. The microwave oven was invented when a scientist was walking past a radar tube and noticed that the chocolate bar in his pocket melted.
5. Ink-jet printers were made possible when an engineer at Canon accidentally put a soldering iron on a pen and superheated the ink.
6. Post-it notes were created when secretaries in 3M discovered a use for non-sticky glue which had been abandoned as a useless idea by managers.
7. Teflon was invented accidentally in 1938 in attempting to make a new CFC refrigerant.
8. Viagra was discovered from treating the side effects of hypertension.
9. Anti-wrinkle creams emerged out of experiments to treat acne with Vitamin A.

But perhaps more than any other modern innovation, the World Wide Web is the most vivid example of how serendipity can change the world. The World Wide Web was just one of many hundreds of transformative ideas to emerge from CERN in the 1980s. CERN is the European Organization for Nuclear Research and remains one of the world's largest and most respected centres for scientific research into fundamental physics. In the 1980s a Technology Transfer Office was created to forge connections between CERN's many global laboratories which were working in relative isolation. Since its formation, this vast new knowledge network has spawned thousands of new serendipitous insights for CERN, releasing an enormous amount of creative energy which had previously gone untapped by the knowledge silos which historically characterized CERN's organizational structure. And it was unexpectedly, out of the collaborative technology networks created by this office, that the World Wide Web was born. As Muller and Becker (2012) point out, CERN's approach to business is a wonderful example of 'planned serendipity' in action. CERN allows its scientists to pursue their own personal passions without concern for short-term commercial goals and its mission is to develop 'ideas' rather than 'products' – many of which are far too advanced for immediate application. However, as with the World Wide Web, while the value of many of the new ideas might not be immediately apparent, CERN's philosophy is that there is a residual value in 'every idea' that only emerges when the timing is right and the world is ready to accept it. CERN's philosophy and approach to business hints at types of strategies which we will explore during the course of his book. As Clegg et al. (2011: 33) point out, in their recent critique of business strategy thinking, the idea that serendipity may play an important role in business strategy is new and 'has been largely excluded from formal strategy theory'. Clegg et al. (2011) argue that there is an element of unpredictability, luck and happenstance in the fabric of all organizations which needs to be better understood for which current models of strategy are inadequate. Indeed, according to Hamel (2002) and Rosenthal (2005), in an increasingly networked, complex and dynamic business world, innovation is much more likely to arise out of serendipity and happenstances than from any formal planning process. So rather than try to eliminate this uncertainty we should accept it as inevitable and learn to manage it.

Making your own luck

Most of us see the concepts of serendipity and luck as synonymous. But all the evidence indicates that luck plays very little part in the lives of successful entrepreneurs. As Mendonca et al.'s (2008) research shows, the concept of luck is an illusion. While on the surface many successful entrepreneurs may seem to repeatedly stumble across good ideas, the key point is that by continually searching for something new, these entrepreneurs are doing something deliberate – even if they happen to find something quite unexpected during the process. In other words, by the very act of looking for something, entrepreneurs instinctively create the 'precipitating conditions' and opportunities for a serendipitous insight to occur.

Lucky people

So it would seem that the old adage is true – 'successful people make their own luck'. However, this is not a random process and Tjan et al.'s (2012) interviews with hundreds of successful entrepreneurs revealed that these so-called 'lucky' people share seven common attributes:

1. Humility
2. Intellectual curiosity
3. Optimism
4. Vulnerability
5. Authenticity
6. Generosity
7. Openness.

According to Tjan et al., these attributes make leaders more 'attractive' or 'magnetic' to others, giving them a greater chance of being exposed to new ideas and insights which might create a new business opportunity. Muir's (2000: 8) research shows that 'lucky' people increase their chances of success by building a 'spider web' of relationships which 'catch' a wide variety of information from different sources. This in turn increases the chances of unexpected encounters with other people who might have a complementary idea.

Success is social

According to Muir, lucky people don't live in straight lines but in a zig-zag. Random interactions, casual conversations and accidental encounters enrich these people's lives suggesting that 'success is social' and depends on relationships rather than facts. This is an idea which challenges the notions of individualism that have traditionally driven Western business teaching, theory and practice. The importance of 'community' is further reinforced by recent research into industry innovation clusters like Silicon Valley. According to Horrigan (2011: 21), people working in firms located in such areas 'come into contact with each other without really understanding what the other is doing, yet they discover that one has the solutions to the other's problems'. These 'communities of practice' expose people to divergent ways of thinking and enable ideas from one discipline to be imported into others. As Mitchell (2012: 8) points out, 'History tells us that when like-minded people come together, great things happen'. For example, the Renaissance in Europe, which produced geniuses like Leonardo da Vinci and Michelangelo, started with the meetings of like-minded artists, scholars and philanthropists in Florence during the late fourteenth century. In Australia, the gold rush in the late nineteenth century drew in thousands of Chinese migrants who eventually established the Chinatowns of Melbourne and Sydney which have acted as incubators of Chinese entrepreneurship ever since. And in the US, the tiny island of Manhattan, which became the home of many European migrants after the Second World War, was the breeding-ground for many of America's modern entrepreneurs in industries like media and advertising.

People are critical

Managers know from experience that while there may be laws and equations that govern the physical laws of nature (such as the law of gravity or thermodynamics), organizations are made up of people who do not behave rationally and predictably. And recent research, which we will discuss in this book, shows that organizations are socio-technical systems where some element of human interaction and unpredictability is inevitable and indeed necessary. As Merton and Barber's (2003) critically acclaimed book *The Travels and Adventures of Serendipity* shows, innovation is a messy blend of inspiration, perspiration, error and happenstance which, despite the rhetoric of scientific method, cannot be perfectly understood and managed. Furthermore, as we have recently seen in the global financial crisis, organizations exist in a highly complex, interconnected and dynamic world which we don't yet understand, where a small change in one remote part of the system can produced immense consequences for those in other parts of the system. Realizing this and confronting traditional notions that we can somehow calculate and control uncertainty is one of the key changes in thinking which managers need to make if they are better to harness the opportunities which lie hidden in the untapped serendipities of the world around them.

Luck is improbable

The concept of luck has also been disproved mathematically. For example, in his book *Struck by Lightning* Jeffrey Rosenthal (2005) argues that all seemingly lucky, random, coincidental and unexplainable events can be explained logically by probability theory. Rosenthal argues that in all systems (natural and manmade) there are patterns that appear to be random events but which are in fact an inherent part of that system. Rosenthal shows through a variety of simple examples and experiments that even the most astounding coincidences are almost inevitable. They only appear to be surprising because of most people's ignorance of how the laws of probability work.

According to Rosenthal, knowing some basic rules of probability allows us to make better decisions and to understand and explain the world around us more clearly. For example, in one revealing experiment, Rosenthal asks blindfolded respondents to place dots randomly onto a piece of paper. Inevitably, as time passes, chance alone ensures that the dots formed clusters and patterns. But this gives the uninformed an impression that the dots are not random at all and indicate some deep and meaningful insights. While this mistake may be harmless in a laboratory setting, it becomes potentially very dangerous when managers and politicians start making decisions on such faulty reasoning. Rosenthal argues that randomness tricks people into seeing patterns and relationships that are not actually there.

This natural 'clumping' effect has been known about for some time and is the basis of the Poisson distribution, first computed by the French mathematician Siméon-Denis Poisson in 1837. Incidentally, this is also the basis of probabilistic risk analysis used in much of business today. But without this knowledge, most people put this natural clumping effect down to luck or bad luck when in fact we are simply seeing a natural phenomenon at work. For example, we might see four car accidents while driving to work on a particular day after having driven for several weeks while seeing no accidents. Most people would naturally attribute the cause of the accidents to something different about that day (perhaps it was raining). But while this might be true, what people might also be seeing is nothing more than Poisson's clumping effect at work. One can imagine many other situations where a similarly misinformed conclusion would be drawn from perfectly natural clumping events such as a spate of murders, a cancer cluster, a spate of plane crashes or numbers drawn from a lottery or accidents across a portfolio of construction projects. Clearly, before jumping to conclusions about such patterns under the pressure to be seen to be doing something to respond, managers and policy-makers should undertake a thorough analysis to get to the underlying facts behind any apparent patterns.

Philosophically, one important question which arises from the above discussion is that if we had enough information, then wouldn't randomness disappear, leaving us with complete certainty about the future? This indeed is the traditional belief which has driven science and the command-and-control models of organizing which have dominated management thinking and practice for so long. But luckily for innovation, these 'deterministic' approaches to management which have sought to eliminate uncertainty and impose universal solutions have been discredited in recent years. As we will show in this book, we now know that no matter how finely one measures a system, no matter how much our technology advances, no matter how much knowledge we acquire, there will always be a certain level of uncertainty and apparent randomness in everything we do or observe. In other words, if you repeated exactly the same experiment many times under exactly the same conditions then nature's inherent randomness might cause a completely different result. This is wonderful news because it is in this uncertainty and unpredictability that the opportunities to innovate lie. A perfectly predictable world would also produce a homogenous world, where there would be little opportunity to differentiate through innovation to achieve a competitive advantage.

There are many more examples where people might attribute what might seem like an extremely unlikely event to luck, when in fact the probability of it occurring was quite high. For example, we are often surprised when we meet people with common acquaintances in the most unexpected places. However, as famously demonstrated by Professor Stanley Milgram in 1967, we are all connected on average by only six degrees of separation, so the likelihood of meeting someone who knows someone you know at any time is actually rather high. To prove this, Milgram mailed packages to random people in the US, requiring them to post them to an unknown target person in Massachusetts via people that they knew on a first name basis. On average, Milgram found that the packages took about two weeks to arrive and had only been posted six times.

The power of collaboration

In proposing a different business model which can capitalize on an increasingly unpredictable, interconnected and dynamic business environment, the aim of this book is to help managers increase the chance for innovation to occur. To this end, this book builds on the work of scholars who argue that serendipity is not accidental, pure chance or pure luck, but is cultivated. As Muller and Becker (2012), show, it is this approach that lies behind the success of what we recognize as some of the world's most well-known and successful companies such as IKEA, 3M, Xerox, Apple, Proctor and Gamble, IBM, Google and Facebook. These companies, including a few leading firms in the construction industry which are discussed later in this book, design their business deliberately to maximize the likelihood of 'chance collisions' between people and knowledge which are normally separated by traditional business practices and knowledge silos. What distinguishes these businesses from others is that they design a certain amount of uncertainty into their business and capitalize on serendipitous insights that emerge out of unexpected collisions between knowledge and people. These organizations actively set up opportunities for serendipity, they put themselves in front of new experiences, are willing to 'go with the flow' and have the vision and adaptive capacity to see and seize opportunities that unexpectedly arise.

Unfortunately, we shall see in this book that the way in which the majority of the construction industry is organized and managed at both project and enterprise level has the effect of reducing the chances for serendipity to occur. And we will also see that this is not at all unique to the construction sector. For example, Clegg et al.'s (2011: xxiv) analysis of business strategy outside construction suggests that traditional strategic thinking, which he argues has emerged "as a macho, testosterone-charged younger brother of economics" provides little insight into how managers might work in this way. This is because it has been largely separated from advances in sociology, organizational theory and political analysis. According to Clegg et al., while good strategy in the past has been seen to arise from the inspirations of great individual leaders, future approaches to strategy will need to be 'constituted and enacted organizationally' and in collaboration with stakeholders. As one of the industry contributors to chapter eight pointed out, Toyota is generally seen as the pin-up company in showing how strong strategic alliances with supply chain partners can transform an industry. Toyota is widely praised for challenging the earlier 'Detroit model' of car manufacturing which promoted a harshly confrontational price cutting environment in its supply chains. As we will see in subsequent chapters, many clients and firms in the construction industry still work on the Detroit model and this represents a significant barrier to innovation.

The idea that success depends on value rather than price and on relationships with others rather than on individual attributes will require a fundamental change in management thinking. Much of what we have been taught about 'good' management comes down to competition, top-down strategy and to individual factors such as talent, attitude, intelligence, education and hard work. So to be able to embrace a new approach to innovation, managers will have to unlearn the lessons of individual achievement that have underpinned the past. They will have to learn the new language of collaboration and to understand the critical role that 'social networks' and 'social capital' play in achieving personal and business success. In sociology, the concept of social capital explains the power and economic benefits derived from one's positioning and quality of relationships within a social network (Burt 2000, 2005). Like other forms of capital (economic, human, intellectual, physical), social capital has an important role to play in increasing productivity (both individual and collective), so do social contacts affect the productivity of individuals and groups. Muir (2000: 24) sums this up nicely when he states that managers will need to shift their paradigm from 'living in little boxes' to 'living in networked societies'. In networked organizations boundaries are more permeable, interactions and actors are more diverse and hierarchies are flatter and more recursive.

This view of future success, which depends more on collaboration rather than competition, is supported by Hagel et al. (2010) in their bestselling book *The Power Of The Pull: How Small Moves, Smartly Made, Can Set Big Things In Motion*. In their seminal book, Hagel et al. argue that new communication technologies are undermining traditional top-down approaches to strategy (the 'push' approach), by connecting employees and empowering them to shape their working lives. In particular, the internet is allowing people to interact with enormous numbers of people who would have been unreachable in the past and social media such as Facebook and LinkedIn are enabling the creation of many new relationships and collaborations which were only dreamt of only five years ago. The implication for managers of these profound changes in the way we communicate is that companies can no longer rely on the stocks of knowledge that they have carefully built up, protected and stored away over many decades. And they can no longer rely on a few old relationships and sources of knowledge that have served them well in the past. The important point being made by Hagel et al. (2012) is that information is more dispersed than ever before and firms must learn new skills to enable them to tap into multiple and more informal sources of information which are often contradictory and constantly changing. While on the downside this appears to be problematic in making it far more difficult for managers to find answers to their questions in a predictable and planned way, there is also an important upside. The vast and interconnected nature of today's social and communication networks means that it is more likely that managers will discover a new opportunity through a chance encounter.

To enable this to happen, Hagel et al. argue that managers must reorganize their lives to increase their chances of unexpectedly bumping into something new. This they argue comes down to one essential attribute – the 'power of the pull' – a magnetism which attracts people and by doing so reveals unexpected serendipities, complementarities and alliances which otherwise remain hidden. To Hagel et al. (2010), the 'power of the pull' enables businesses to find and access people and resources when they need them and it is especially important in a world of depleting human and natural resources. This has been recently confirmed in the construction sector by Seneghi and Loosemore (2012), whose research into 'employers of choice' indicates that in a world of critical skills shortages and an ageing population, employers that fail to provide excellent workplaces will lose their most talented workers to more 'magnetic' organizations that provide better chances for learning and growth. For Hagel et al. (2010), one of the critical defining factors that distinguish innovative companies is their socially imbedded management practices which build and harness loosely coupled business networks to provide scale without inertia. These business networks drive learning, capability building and performance improvement by creating the conditions to enable people to learn faster. In the most successful firms, these networks are often based on deep and trusting personal relationships and their members are drawn from diverse backgrounds. Like many pieces of research before them, Hagel et al. confirm that diversity is critical for innovation because it delivers the creative tension from which many innovations are born.

Despite the growing importance of collaboration to innovation, recent research by Palangkaraya et al. (2010) indicates that the legacy of modernism in management has ensured that the firms are struggling to leverage the opportunities in their business networks and supply chains. Their analysis of patent applicants in the resources, services and manufacturing sector in Australia between 1986 and 2005 shows that the number one barrier to innovation was 'finding a partner' and the second was 'lack of complementarities'. Recent research suggest that the same is true in the construction sectors. As Tennant and Fernie (2011) point out, while over the last decade principles of supply chain management have been promoted extensively within the construction sector, to date the focus has been on tier-one and tier-two relationships and has rarely extended into downstream supply chain relationships, leaving the aim of fully integrated supply chains an aspiration rather than a reality. There is little evidence to indicate that the problems of supply chain management in construction highlighted by earlier research into innovation by Tatum (1987), Lamborde and Sanvido (1994) and Gann (2000) have been resolved to any large extent. They, like many others since, argued that the fragmentation and one-sided management of risk in the construction supply chain separates innovators from the potential benefits of their innovations and is one of the greatest barriers to innovation in the construction sector.

The power of interdisciplinarity

Camillus (2008) argues that the world's new and poorly understood complexities present managers with new types of 'wicked problems' which are messy and difficult to resolve using traditional command-and-control structures. While in the past managers may have been able to tackle most problems in advance and work in relative isolation to develop solutions, these problems will require managers to 'learn as they go' and to do this in more collaborative environments where business, communities and governments work in new organizational configurations which are likely to challenge traditional competitive relationships and perceptions. According to Locke and Spender (2011), it is in newly forged relationships and at the intersection of different knowledge domains that firms will find the solutions to today's new challenges. Many of the world's most innovative firms recognize that breakthrough ideas will most often occur when they bring concepts from one field into a new and unfamiliar territory. For example, of the approximately 6000 people Google employed in 2013, 4000 to 5000 were from the humanities and creative arts (Cai 2013). IBM is also recruiting classics graduates and many leaders of the corporate world have a non-business background. David Thodey, CEO of Testra, was an English and anthropology graduate, Gail Kelly, CEO of Westpac Bank and the first female chief executive of a major bank, taught Latin at an exclusive boys' school in Zimbabwe and Kim Williams, head of Rupert Murdoch's empire in Australia, was trained as a classical composer.

Evidence to demonstrate the power of interdisciplinarity in offering new insights and solutions to seemingly intransigent and perplexing problems abound. Charles Darwin was a geologist when he proposed the theory of evolution and it was an astronomer who finally explained what happened to the dinosaurs. In the corporate world, Nestlé developed the idea of its flavoured drinking straws by adapting similar products used in hospitals to administer bad tasting drugs to children. And as long ago as 1966, an architect in Zimbabwe borrowed ideas from observations of how termites ventilate their nests to design an energy-efficient shopping centre in central Harare in Zimbabwe. Designed to be ventilated and cooled by entirely natural means, this was probably the first building in the world to use natural cooling to this level of sophistication. Today, biomimicry is one of the most exciting examples of how disciplines can collide to drive innovation in the construction sector. For example, Elizabeth Demaray, a professor of fine arts at Rutgers-Camden, has developed a lichen slurry that can be painted onto buildings. 'Lichen can live where plants cannot, and it can thrive without human intervention or maintenance. Also, anyone can "plant" it – all you have to do is smear it on the outside of your building', Demaray says (Mims 2011). Lichen is a unique collaboration of fungi and algae, in which the fungus provides structure and protection to the algae, which feeds itself and its host through photosynthesis. The value of lichen is that it can inhabit surfaces and climates where no other living thing can get a toehold and its light colour and respirational properties help cool anything it grows on. It is not hard to imagine how the sustainable city of the future may be cooled in this way. Indeed, a new building on 14th Street in New York City is contemplating covering its entire facade with it. A lichen-encrusted building would help keep itself and the city around it remain cool. While ideas like these may be in their early stages, the key lesson that we can take away from this type of innovation is that collaboration is key. Diverse teams create far more ideas than homogenous teams.

The risks and opportunities of interdependence

Recent global events such as terrorism and the global financial crisis have highlighted like never before the intricate and contagious ways in which modern business, society and governments are interdependent. As the World Economic Forum's *Global Risks Report 2013* pointed out, new inter-related and often invisible global interdependencies have emerged in recent years to pose enormous levels of systemic and contagious risk to global governments, businesses and communities around the world (WEF 2013). Exacerbated by the breakup of traditionally vertically integrated industries and by increasing regulatory and technological complexity, we now face a world with multiple and potentially cascading interdependencies and uncertainties which did not exist in the past.

The consequence of this intricate interdependence is that an isolated problem in one part of the world is now capable of spreading and magnifying in an uncontrolled manner to seemingly remote and unrelated businesses and societies in other parts of the world. These increasingly frequent scenarios are reminiscent of the 'butterfly effect', an imaginary notion where a butterfly flapping its wings in one part of the world could conceivably cause a tornado in another. While this effect is an exaggerated hypothetical example of chaos theory first proposed by American meteorologist Edward Lorenz and later popularized in the 2004 Hollywood movie of the same name, it is becoming increasingly accepted that in today's business world, small initial changes in apparently innocuous places can, over time, lead to large unforeseen consequences in another.

So it would seem that we may be entering a new era of instability which holds both enormous new opportunities and risks for business for which most businesses are completely unprepared. As Rosenthal (2005) argues, we have two options in the face of this inevitable randomness and uncertainty. We can either let it get the better of us or we can acknowledge that it is inevitable and healthy and learn to understand and embrace it. In essence, this is the focus of this book. First it argues that we need new ways of thinking to harness these opportunities. Second, it argues that a better understanding of the innovation process holds the key to this, whether you are an architect, engineer, small subcontractor, manufacturer, supplier, consultant or a major multinational developer. Contrary to what is often assumed, innovation is not confined to the creative industries like architecture, art and design. As demonstrated by Samson's (2011) cross-sector research into innovative companies such as Lonely Planet, Toyota, Microsoft, Newcrest Mines, Strechtex, Textor Technologies, The Speciality Group, Ferguson Plarre, GPC Electronics and GRL Mobile, the defining characteristic of innovative companies was not their size or function, but a 'systematic innovation capability' which consistently delivers innovations which add business value. Systemically innovative firms back up their intentions with resources, measure innovation outputs, encourage staff to contribute towards it and reward them for doing so. In these companies, innovation is deeply embedded into the mindset and culture of the business and there are close collaborative relationships with supply and demand chain partners to achieve win-win innovation outcomes. This in turn requires a number of generic attributes such as a clear strategic focus on innovation, a tolerance for experimentation and risk and strong leadership which makes innovation a priority.

Unfortunately, according to a recent panel discussion of Australia's business, community and political leaders, while there is a depth of management talent in industry, there is a dearth of exceptional leadership (Fox and Hooper 2011). According to Fox and Hooper (2011: 22), business has been largely disengaged from contemporary social challenges such as climate change and environmental degradation and 'missing in action as citizens'. Rather than simply speaking on behalf of their corporation and the narrow short-term business interests of shareholders, business people who speak as citizens and take responsibility for the wider community in which they do business are 'pretty much unheard of' and 'almost entirely absent'.

The drivers of innovation

The drivers of innovation are of course constantly changing and include increasing competition, shareholder demands for growth, regulatory requirements, changing customer expectations, innovative business partners or simply having to resolve a problem. Sometimes the drivers of innovation are also counter-intuitive. For example, a recent Ernst and Young survey of business leaders conducted in the midst of the global financial crisis found that a majority of entrepreneurs saw economic downturns as the perfect time to innovate. The research notes that many of the world's most respected companies were born during a recession and while instinct might cause most people to retreat, withdraw and cut in a downturn, these difficult periods force people to search for better ways of doing things through creative and unconventional approaches to doing business. As one respondent said, innovating during the global financial crisis 'isn't for the faint hearted, but there are few golden moments in history like this one to dominate your industry and your competitors' (Ernst and Young 2009).

The focus for innovation can also vary enormously between firms. Some organizations will focus on improvements to productivity and efficiency while others will look for advances in product/service quality; the development and creation of new markets; extension of existing product/service ranges; or reduced environmental and social impacts. Manley and Rose's (2009) analysis of innovation in the construction sector found that while cost cutting is often assumed to be the main focus of innovation, ambitions around improving sustainability, energy efficiency, environmental improvement and lifestyles are far more prevalent. This reflects growing agreement that while innovation is essential to economic prosperity it has become even more critical to social progress, sustainable development and the battle against unemployment, poverty and inequality around the world (Le Masson et al. 2010).

Pro-innovation bias

With all the recent hype surrounding the topic of innovation, it is far too easy to uncritically embrace the idea without considering the negative social, economic and ecological costs it may have to communities. While innovation can no doubt play an important role in meeting the world's future challenges, any suggestion that a new idea should be uncritically adopted by 'all' members of society is dangerous and ignores its potential downsides (externalities) of innovation. As Rogers' (1995: 104) analysis of innovation rhetoric revealed, most innovation research and literature is uncritical and heavily laden with positive value – 'Innovativeness, like efficiency, is a characteristic we want all organisms to possess'. Rogers argues that the potential downsides of innovation are written about far less frequently than its benefits. He argues that this is because much of the literature in this area is produced or funded by change agencies which have vested interests in seeing certain innovations adopted. These biases are then injected in the work of the researchers and consultants whose work they sponsor.

A good illustration of this in the construction sector are Public Private Partnerships (PPPs) or Private Finance Initiative (PFI) projects. PPPs are a process innovation introduced for reasons of financial expediency around the world as the answer to public infrastructure provision in a resource-constrained public sector. Flyvbjerg (2005) showed how significant optimism bias and strategic misrepresentation is built into the feasibility studies of these projects which are often signed off with an inadequate appreciation of the risks and externalities involved (both actual and publically perceived). The result of this bias, despite many success stories, has been a number of high-profile failures which have caused public perceptions of these projects to become particularly negative and hard to change. This has been especially true in road and bridge infrastructure where the public has boycotted numerous road, rail and tunnel projects in protest against what they 'perceive' to be poor consultation, unacceptable social impacts and high tolls and road changes which force them to use the new infrastructure. In many cases, overly optimistic patronage estimates have required the public to compensate consortiums for revenue shortfalls and even bankruptcy to the tune of hundreds of millions of dollars. The health sector has also proved contentious with many clinicians arguing that PPP hospital projects are driven by private rather than public interests and compromise health service outcomes. Writing in the *British Medical Journal*, clinicians described PPP projects in the health sector as 'Perfidious Financial Idiocy' (Smith 1999), perceiving PPPs to be moving priorities for primary care into corporate hands. These perceptions of externalities have been reinforced recently as UK taxpayers have had to fund a £1.5bn bail-out of twenty-one PFI hospitals which were at 'severe risk' of failing patients (Mason 2012). What is clear from these projects is that public perceptions matter in any innovation. As Sharpe (2004: 8) noted, while PPP projects might offer creative solutions to public infrastructure needs, at the core is a complex web of relationships among bureaucrats, politicians, media, employees, general public (local, national and sometimes global), labour and special interest groups – 'Any PPP lives or dies on its reputation with these people'. This recognition that society, communities and 'normal' people have a critical role to play in the innovation process is a recurring and important theme throughout this book.

More recently, Green (1998, 2006, 2009), Ness (2010) and Ness and Green (2012) have shown how pro-innovation bias is also clearly evident in construction industry reform agendas around the world. They argue that governments and businesses in the sector too readily adopt management 'recipes' which are often imported from manufacturing, without understanding the significant risks that accompany them. The concepts of lean construction, business process re-engineering, benchmarking and the UK government's 'Respect for People' initiative come in for special criticism for masking the harsh realities of working life in the construction industry and for exacerbating the causes which fuel the problems in the first place. Loosemore and Phua's (2011) research also revealed a significant level of pro-innovation bias around corporate social responsibility in the construction sector. They pointed to accumulating evidence that the consequences of making unsubstantiated claims and investment decisions on the basis of biased evidence in this area is highly counterproductive to the finances and reputations of the organizations involved.

Given the above, it is critical that any claims about the benefits of innovation are based in fact not fiction and that companies also acknowledge the potential downsides of the innovation process. While this may be uncomfortable, by doing this they will be in a far better position to ensure that the outcomes of the innovation process are positive and achieve their stated aims.

The pathology of innovation

As a counterpoint to the pro-innovation bias which characterizes much of the innovation literature, it is important to recognize that innovation also has a dark side. Failure is a recurring theme in the stories of many of the world's influential innovators and as much, if not more, can be learnt from innovation failures as it can from those who succeed in the process. For example, despite becoming a global phenomenon, Facebook did not grow without its problems and failures. Few people realize that Mark Zuckerberg lost US$174 million in his first two years of operation. And Sir James Dyson, the inventor of the world's first bagless vacuum cleaner, openly admits that 'I made 5,127 prototypes of the vacuum before I got it right. There were 5,126 failures. But I learnt from each one'. Dyson's story is typical of many successful entrepreneurs. His early experiences in transferring his ideas into commercial reality were punctuated by continuous failure and years of fighting, failed business pitches and rejections by big players. As he points out, 'You need dogged determination and stubbornness and every time you're knocked down, bounce back up', particularly in the early years. Now worth an estimated $1.45 billion and selling products in over 50 countries. Dyson argues that many Western countries, like the UK, have become too scared to take risks. Dyson dispels the myth that successful entrepreneurs move from one success to the next and that all you need is one good idea and you will make your fortune. Rather, he argues, successful entrepreneurs start with a problem that they are determined to solve and then build and test solutions and prototypes – hundreds or perhaps even thousands of them. Often, he argues, the final problem or solution doesn't look anything like the original one you started off with. The important thing is the journey and the learning that occurs during that process.

Much research supports the anecdotal insights above, confirming that risk and failure are an inevitable and important part of the innovation process. According to March (1999) this dark side of innovation is driven by four pervasive features of organizational life:

1. the inevitable obsolescence and eventual failure of new ideas and technologies leads to the search for new ones which in turn fail, leading to more searching and so on;
2. the fact that most new ideas are bad ones and need refinement;
3. there is often significant optimism bias in an organization's aspirations which can blind organizations to the risks of failure and;
4. while the process of innovation creates excitement, energy and inspiration in some quarters, it will conversely disturb existing power balances, creating fear, uncertainty and frustration in others. That is, the innovation process inevitably creates winners and losers.

The value of this work is clear. If we are to fully understand the innovation process we must take a balanced view from a range of perspectives and not from just the vested interests of those who are driving it. For example, a new idea might render obsolete existing products and services and potential businesses and jobs in other parts of a business or an economy. Furthermore, to ignore and hide the almost inevitable failures associated with the innovation process is dangerous. Not only can this mask potential risks and externalities associated with innovation but it can also hide important insights and lessons which can be gained to refine and improve future innovations. In addition, refusing to acknowledge failure can obscure the factors that influence the public's rejection or discontinuance of innovations (for example due to unacceptable negative consequences to social systems), how we can prevent the diffusion of bad innovations (that can harm society) and the process of adaptation and reinvention undertaken by many users as they mould an idea to suit their exact needs (which naturally occurs during the innovation process).

The externalities of innovation

The realization that new ideas create winners and losers and affects different stakeholders in different ways is important in understanding the potential externalities of the innovation process. As a recent study by Ernst and Young (2009) showed, the market leaders of today are not necessarily the market leaders of tomorrow. For example, over the last five years the Russell 3000 (the top 3000 public companies by market capitalization) has experienced 59 per cent turnover, the Global Forbes 200 51 per cent, the HDAX in Germany 50 per cent, the FTSE in the UK 50 per cent, the KOSPI in South Korea 49 per cent and the Bombay 200 in India 90 per cent. Long ago, political scientist Joseph Schumpeter, one of the most influential economists of the twentieth century, recognized this phenomenon and called it 'creative destruction' (the demise of old businesses by the birth of new ones) (Schumpeter 1947: 18). Schumpeter argued that this process represents the foundation of capitalism and market economies. Building on the work of Karl Marx, Schumpeter developed the theory of economic innovation and business cycles, identifying innovation as the critical dimension of economic change. In Schumpeter's vision of capitalism, technological innovation often creates temporary monopolies, allowing abnormal profits that would soon be competed away by rivals and imitators. Schumpeter argued that these temporary monopolies and their associated super profits were necessary to provide the incentive for firms to develop new products and processes. His work sought to prove that innovation-originated market power could provide better results than price competition in the long term. To Schumpeter, innovative entry by entrepreneurs is the necessary disruptive force that sustains economic growth, even if it destroys the value of established companies that enjoyed some degree of monopoly power derived from previous innovations.

While destruction is a fundamental consequence of the innovation process, Aghion and Howitt (1998) argue that it is not always as positive as the term 'creative' implies. Indeed, innovation can involve considerable waste of human and physical resources through the premature abandonment of ideas, products and services which may still have some residual value to society. In reality, old products can and do exist alongside new products (if new products are introduced too early) and new products tend to be imperfect substitutes for old products (therefore prolonging old products' life). Nevertheless, the idea of creative destruction does raise some important moral and economic questions about the 'value' of innovation and whether there is an optimum level, timing or pace of creative destruction from society's perspective. If companies get the timing, pace or level of innovation wrong, then they can expect to incur the wrath of increasingly informed, empowered and vocal constituencies.

The spill-over effects of innovation

Because the concept of innovation originated in the field of economics, it has traditionally been concerned with the commercial perspective of individual entrepreneurs, firms or governments seeking to add to their own wealth, power and prestige. However, today the concept of innovation has been broadened to include social advantage as well as economic benefit. It is now accepted that innovations have important, significant and often unintended positive 'spillover' benefits for societies which were not targeted in the original innovation – both social *and* economic. While many firms try to minimize these spillover effects through mechanisms like patents, to maximize their individual return-on-investment, Baumol (2010) argues that innovation spillovers have been responsible for most of the Western world's prosperity over recent centuries. Baumol estimates that the spillover effect from innovation in the US conservatively accounts for more than half of the growth in current GDP. He also argues that in a world where almost all of the world's major technological developments occur in just twenty-five dominant countries, this spillover effect is more important than ever in addressing global poverty and economic inequality between developing and developed countries. Baumol's analysis shows how most of the innovations experienced in developing nations do not come from that country's own research and development but from the investments transferred or even copied from more wealthy countries.

Innovation – an idea without a theory

Until recently, despite Schumpeter's early work, the subject of innovation was not considered a subject worthy of serious empirical study. Consequently, innovation remains almost completely excluded from modern theories of the firm and we only have a rudimentary understanding of the factors that might influence an organization's innovative capacity. Baumol argues that the main reason that innovation has not been incorporated into the modern theory of the firm is that sustained market equilibrium (the foundation of most economic models) is not tolerated by entrepreneurs. As Schumpeter (1936) argued, by definition the task of the entrepreneur is to search for opportunities to create and then exploit market disequilibrium. Then, by taking advantage of the opportunity (and by encouraging others to copy the new idea), the entrepreneur provides the force to move the market back to equilibrium, and so on.

While a theory of innovation does not yet exist in mainstream economics and management, there is even further to go in conceptualizing innovation in project-based industries like construction. Here, the innovation process may be quite different from models developed in the field of manufacturing. Widen (2006) recognized this when he argued that, in contrast to mass production industries where new ideas can flow relatively easily through vertically integrated organizations, new ideas in construction are applied in individual projects with 'hard' boundaries. This not only acts to prevent their diffusion but it also prevents the amortization of development costs, risks and benefits across different individual projects. In reality new ideas are continuously being developed on many complex construction projects to overcome the unique challenges each pose. However, without diffusion to other projects, these activities do not fit neatly within definitions and models of innovation being developed in manufacturing. Furthermore, in contrast to much of the manufacturing sector, the products of construction are delivered by temporary, transient and highly fragmented project organizations involving a multitude of subcontractors, consultants and suppliers arranged into long and complex supply chains with complex risk structures and often conflicting interests. Since the benefits and risks associated with innovations inevitably affect more than one organization in this type of structure, it is more difficult for one organization to take responsibility for it and for the benefits to be spread between project participants. Furthermore, because most projects last a relatively short period of time, are often highly unique (apart from in sectors like housing and industrial construction) and because the volume of production (which is essentially small-batch) is not large enough to cover the development costs of innovation, the costs of innovation tend to be higher and carried by individual projects. On top of these problematic characteristics, there are many other cultural, institutional and organizational barriers to innovation which are tied up in hundreds of years of tradition, professional differentiation and established ways of working in the construction industry. In long-established industries like construction, innovators are faced with the daunting task of challenging hard-wired and resistant 'path-dependencies' upon which people, organizations and institutions depend for their survival. As the late Abraham Kaplan pointed out, the more we know about how to do something, the harder it is to learn how to do it differently (Kaplan 2004). The well-known American psychologist Martin Seligman coined the term 'learned helplessness' to describe this trained incapacity to innovate.

Conclusion

The aim of this chapter was to introduce the basic principles of innovation and explore some of the myths that surround the process in the construction industry. Putting the case for a more informed and enlightened approach to the subject, it has questioned the future traditional mechanistic approaches which assume the world is stable and linear. In contrast, a more organic approach is advocated which is able to harness the untapped serendipitous opportunities presented by an increasingly dynamic, uncertain and interconnected world. This idea is explored in much more detail in subsequent chapters. However, it is evident already that this will involve a focus on consultation, collaboration, diversity and understanding the power of networks in building social capital. The next chapter takes these formative ideas forward and discusses the process by which managers integrate them into a business strategy.

2 The innovation process

> If you do not expect the unexpected you will not find it, for it is not to be reached by search or trail.
>
> Heraclitus

Introduction

Innovation is the process by which organizations successfully transform new ideas into improved products, services or processes, in order to advance, compete and differentiate themselves in their marketplace (Baregheh et al. 2009). For innovation to occur, something more than a creative insight is required. Innovation involves acting on creative ideas to bring them to reality and implementing them to make a tangible difference in the community in which the innovation occurs.

As a result of Schumpeter's (1936) formative work, five broad types of innovation are widely recognized:

1. Product innovation – creation of new or improved goods and services that are launched onto a market
2. Process innovation – changes in the way that goods and services are produced which includes introduction of new procurement methods and production technologies
3. Organizational innovation – changes in the architecture of firms, management structures, roles and responsibilities, corporate governance, financial systems, remuneration systems, industrial relations reform, supply chain reform etc.
4. Input innovation – improved ways of sourcing supplies, raw materials and intermediate goods and services
5. Market innovation – opening up of new market opportunities.

While these definitions and categorizations of innovation are useful, they also raise some important questions. For example, the concept of 'success', which is so central to innovation, means different things to different stakeholders. So can an idea which benefits one group and disadvantages another legitimately be classified as an innovation? Sexton et al. (2008) argue that to be true innovation its benefits must be shared between all stakeholders.

The benefits of innovation can also be realized over different timescales. So does a new idea which might not yet have an application still classify as an innovation? For example, Einstein first theorized about lasers in 1917 but it was not until Charles Townes' work in the 1950s that laser beams were created and later in 1960 that the first functioning laser was developed at Hughes Research laboratories in Malibu, California, by Theodore H. Maiman.

Another contentious question is whether a company has to be the first to introduce a new idea to be legitimately labelled an innovator. For example, NESTA (2007) defines an innovative organization as one which 'first' develops a new idea. Organizations that follow are classed as adopters of innovations and those which supply ideas to others such as a trade association as transmitters of innovations.

Finally, what classifies as a 'new' idea and how radical does it need to be to be classed as an innovation? This is important because in industries like construction, most innovations are incremental improvements to products and processes which arise out of having to solve day-to-day problems. For example, does the incorporation of safety handrail lugs into precast bridge components to prevent falls during construction carry the same weight as radical transformative innovations like the creation of the Ipad? And is the answer to this question context dependent? Because a major innovation for a small firm might be a minor innovation for a big firm. Dodgson and Gann (2010) argue that in reality there are different 'levels' of innovation. These range from continuous 'incremental improvements' in products or services (which seek to acquire a larger slice of the pie), to 'radical changes' in the nature of products and services (which seek to enlarge the pie) to 'occasional transformational' innovations (that create entirely new pies). In contrast, Hamel (2002) argues that it is only the radical, disruptive 'rule-busting' innovation that can create lasting and substantial benefits for stakeholders. These 'nontrivial' innovations rip up the rule book and dramatically change and dislocate the direction of an industry or society. These radical transformative innovations leave competitors scrambling to catch up, they affect entire industries and are multi-dimensional in the changes they produce. Contrary to what most people would assume, Hamel (2002) argues that these types of innovation are not beyond the realm of small organizations and do not need huge investments in R&D which can only be provided by large organizations. As the late Steve Jobs, the founder of Apple said

> Innovation has nothing to do with how many R&D dollars you have. When Apple came up with the Mac, IBM was spending at least 100 times more on R&D. It's not about money. It's about the people you have, how you're led, and how much you get it.

What defines a disruptive innovation is not the amount of resources invested in it but the extent to which it departs from industry norms. That is, the extent that it renders existing business models obsolete, changes the basis of competition in an industry and produces sustainable competitive advantage by changing the way a whole industry works.

Hidden innovation in the construction industry

Historically, most developed countries' innovation policies and definitions of innovation derive from the Oslo Manual (OECD 2005) and Frascati Manual (OECD 2002) respectively. The Oslo manual takes a narrow neo Schumpeterian view of innovation focusing on R&D based technological product and process innovation in the business sector. The Frascati manual defines three activities that are considered to constitute R&D. These include:

1. Basic research (experimental or theoretical work undertaken to acquire new knowledge without any particular application)
2. Applied research (original investigation undertaken to acquire new knowledge directed towards a particular application)
3. Experimental research (systematic work derived from research, existing knowledge or practical experience aimed at developing new materials, products systems or services).

Both the Oslo and Frascati manuals have been criticized for favouring traditional R&D-based notions of innovation. It has been argued that this limits their value to creative industries such as architecture where social sciences, arts and humanities research is more relevant than traditional experimental science (Taylor 2005). Furthermore, it has been argued that the Oslo and Frascati definitions favour the type of innovation that occurs in high-technology product-based manufacturing type industries and undervalues the type of innovation that tends to occur in low-technology, service-based and creative industries like construction. Since most governments around the world use the Oslo and Frascati definitions as the basis of their innovation indexes, this ensures that much of the innovation that occurs in the construction sector goes unnoticed and is invisible to traditional metrics and measurement frameworks used to rank industries. It is for this reason that the construction sector often appears to underperform in its innovation output compared to other industries like manufacturing. And it is for this reason that the construction industry tends to be positioned outside national innovation agendas which tend to support the type of traditional R&D-based innovations which occur within the manufacturing sector.

It is also argued that traditional approaches to innovation informed by the Oslo and Frascati frameworks fail to understand recent changes in modern business models brought about by the growth of new creative economies, by the digital revolution and by the increasing interplay between technology-based and creative industries. As Taylor (2005: 10) points out, we 'no longer live in a world of two cultures where art and science are artificially polarized by outdated frameworks of classification'. The boundaries between service-based and manufacturing-based industries are blurring too and it would seem that the foundations of many national innovation systems are too narrowly defined to drive a modern knowledge economy. Cutler (2008) recognized this problem in his review of Australia's national innovation system, arguing that traditional government measures of technical innovation significantly underestimate the amount of innovation occurring in many industries. Cutler also recognized that formal R&D accounts for only one-third of total business expenditure on innovation and that innovation occurs in many different ways. In the UK, NESTA (2007) and Abbot et al. (2007) also highlighted this issue, arguing that there are fundamental differences between traditional science-based measures of innovation (patents and R&D expenditure etc.) and those needed to detect the type of innovation that occurs in service-based industries like construction. NESTA (2007) and Abbot et al. (2007) called the type of innovation which occurs in construction 'hidden innovation' because it is hidden by traditional innovation measurement frameworks indices. NESTA also classified four types of hidden innovation from an analysis of six industries which typically perform badly on traditional measures of innovation such as oil, retail banking, construction, legal aid services, education and offender rehabilitation (see Table 2.1).

Table 2.1 Types of 'hidden' innovation

Type of innovation	*Definition*
Type 1	Identical or similar innovations to traditional scientific innovation but excluded from traditional measurement (e.g. new offsite construction technologies developed with manufacturers and suppliers).
Type 2	Innovations without a major scientific or technological basis (e.g. new business structures or models such as PPPs or alliances).
Type 3	Innovation created from novel combinations of new and existing technologies and processes (e.g. business information modeling software for designing buildings more sustainably or safely).
Type 4	Incremental every day locally developed small-local scale innovations that are normally under the radar which rely on tacit knowledge and learning by doing (e.g. novel design or construction solutions developed in response to site-based problems and generated collaboratively within multidisciplinary design and construction teams).

(Adapted from NESTA 2007)

NESTA (2007) argued that hidden innovations are different in a number of important ways to the types of innovation detected by the traditional Oslo and Frascati definitions:

1. They are not driven by large, formalized, pre-planned, laboratory-based R&D programs but by the necessity of having to respond to day-to-day problems in a real life setting (such as a construction project).
2. They do not generally take the form of new technologies developed through the application of knowledge in the physical sciences. Rather, they tend to arise from applying ideas from the social sciences (traditionally excluded from innovation metrics), from adapting and borrowing ideas from elsewhere (reinvention) and by applying them to new problematic contexts.
3. They do not tend to occur within the confines of one single organization, as they do in many large multinational R&D-based organizations. Instead hidden innovation tends to happen spontaneously and often serendipitously in collaborative networks of manufacturers, consultants, suppliers, subcontractors, designers and clients and often in response to unexpected and novel problems that arise on specific projects.
4. They are less sensitive than traditional innovations to policy framework conditions such as government incentives, taxation policy, skills and regulations and wider political conditions. Instead, high-tech suppliers and intermediaries tend to play a much larger role in stimulating, developing and diffusing this type of innovation than governments.

NESTA argue that the traditional R&D view of innovation is not only inappropriate for the construction sector but is an outdated linear 'pipeline' model of innovation which fails to reflect the type of innovation happening in the burgeoning service-based industries that now dominate many developed countries. It is clear that new innovation metrics are required to detect the real level of innovation that is happening in these low technology intensive sectors. As the ancient Greek philosopher Socrates argued, something does not have to be tangible and 'graspable' to be real. People he argued, who only see with their eyes are blind and that inspiration and imagination are only available to those who use all their senses.

To address this significant problem, in 2001 Latin America and the Caribbean countries created the Bogota Manual which has a much broader definition of innovation. However, there has been precious little work in developing metrics to measure this type of innovation and much more work still needs to be done. To this end, emerging discourses on social innovation, user innovation, open innovation and democratized innovation are useful. They represent the emerging view that innovation comes from many sources, not just the research laboratory, and that there is equal merit in innovating for the betterment of society as there is for advancing an organization's commercial interests. We will discuss these ideas in more detail as we move through this book.

Imitation, invention and imagination

Conversations about innovation can often become confused by the related concepts of imitation, invention and imagination. There is a tendency for people to use these interchangeably whereas, like all words in the English language, they have developed their own precise meanings and place in the contemporary innovation lexicon.

Imitation

While the notion of imitation is now widely seen as inferior to innovation, it was for a long time accepted as an equally respectable pursuit. As vividly illustrated during the 1970s by the booming Japanese industry, through imitation, old ideas such as Western cars and electronic equipment could be copied, refined and improved with significant success. In response to this, previously complacent Western companies did the same, dismantling Japanese products and imitating their components to improve on their own old standards of production. For example, Camp (1989) notes that when General Motors and Ford dismantled the much smoother Japanese gear boxes in the 1970s they discovered that the Japanese were working toward much smaller tolerances than those set in the American goods they copied. This in turn led to new improved standards and regulations in the American automobile industry. So while the concept of imitation is often seen as inferior to innovation, and while it may not be a form of 'primary' innovation, it is nonetheless an important form of 'secondary' innovation which can lead to important incremental improvements in existing products and services.

Invention

Originating in medieval European language, the term invention was for a long time closely associated with the origins of science and interpreted as a process of discovery (finding new knowledge) as well as making new artefacts. During the Renaissance period, there were few distinctions between artists, scientists and inventors. For example, Leonardo da Vinci was a painter, sculptor, architect, musician, scientist, mathematician and engineer, who perhaps more than that of any other figure epitomized the archetypal 'Renaissance Man'. A person of insatiable curiosity and endless inventive imagination, Leonardo is today primarily known as a painter (producing masterpieces such as the Mona Lisa and The Last Supper). However, his notebooks also contain scientific diagrams which show his technological ingenuity and he is still credited with many important discoveries in the fields of hydrology, civil engineering and optics. Leonardo conceptualized a helicopter, a tank, concentrated solar power, the calculator, the double-hulled boat and a rudimentary theory of plate tectonics. Although few of his designs were constructed or were even feasible during his lifetime, some of his smaller inventions, such as an automated bobbin winder and a machine for testing the tensile strength of wire, did eventually come to fruition. However, since he did not publish any of his ideas or findings, they had no direct influence on later development of science.

Despite these early links with the arts, over time the term invention became associated with technology and science. This transition gathered special momentum during the industrial revolution as engineers rose to prominence as the great inventors of their time, creating major leaps forward in social and economic infrastructure through the creation of tunnels, bridges and railways, and latterly electrical systems. Around this time the patent system also developed as did the process of scientific publication for protecting inventions and the huge financial rewards which arose from their commercialization.

Imagination

The concept of imagination has had a turbulent history. For example, Plato rejected the idea of imagination and argued that artists and poets must be banned from society. Aristotle had more respect for artists and realized that artists tried to represent the human experience through universal metaphors and mental images drawn from their imagination. Aristotle realized that throughout history, oral cultures depended on imagination to pass on their myths, stories, songs and dances to the next generation. Indeed, even today some ancient cultures such as Australian Aborigines (through their concept of Dreamtime) still treat the imagined world as having equal validity to the real world. However, most modern societies continue to treat human imagination as a negative attribute and children accused of 'day dreaming' are normally disciplined for wasting their time. Yet it was Albert Einstein, the world's most famous innovator, who famously said that 'Imagination . . . is more important than knowledge. Knowledge is limited. Imagination encircles the world' (Viereck 1929).

Sadly, with the development of writing and science and with the spread of formal education, the value of this form of thinking has been largely forgotten. Scientists have always had a negative view of imagination because it cannot be measured or tested by facts and scientific method. Nevertheless, while science has struggled in accepting the value of imagination, the field of literature and visual arts has made widespread use of the idea, arguing that it is central to the creative process. The concept of imagination is the unique capacity of humans to think of things in terms of possibilities rather than facts (Egan 1992). To psychologists imagination is a form of 'counterfactual thinking' and 'image formation' which takes place without anyone else's knowledge and which involves the creation of new ideas that cannot be perceived through normal sight, hearing, or other senses. The concept of the imagination is complex and has close connections with both memory and emotion. It helps people visualize things which might not be possible in reality and is said by psychologists to be the fundamental cognitive faculty which helps people learn and make sense of the world and move beyond reality to what 'might be possible'. For this reason, it is a very important human faculty in developing an innovation capability.

A very brief history of innovation

The first theories of innovation evolved from the work of French sociologist Gabriel Tarde in the late nineteenth century. To Tarde, innovation was not the result of one lone inventor (who was a hero or genius) but was a social process involving numerous actors. According to Tarde, after being introduced, most innovations are met with opposition and resistance which eventually gives way to acceptance and ultimately imitation and refinement. Innovation was seen to be an evolutionary cumulative process which arose out of lengthy refinements of ideas over long periods of time and from the combination of diverse knowledge from the fields of design, science, materials, method, capital, skills and management. Without an 'inventor', innovation was not possible but individual inventors existed in a social system and had to work within the limitations it imposed.

Over time, economists also contributed to theories of innovation by rejecting the notion that a new idea could be an innovation if it didn't have a market value. Economists defined innovation as the industrial commercialization of technological invention. To economists, the creative abilities of inventors were framed in the context of organizational service, industrial development and economic growth. This led to the development of more contemporary theories around the relationships between innovation and entrepreneurship, and it is to these ideas that we now turn.

Innovation and entrepreneurship

The word 'entrepreneur' has its origins in French and literally means taking the initiative to bridge – 'entre' (between) and 'preneur' (buyer). As we pointed out in chapter one, entrepreneurs play a critical role in the innovation process by creating the critical link between invention and application. Joseph Schumpeter's formative work on innovative entrepreneurship in the 1930s and 1940s is widely credited as offering the beginnings of an operational theory to explain this link (Schumpeter 1936, 1947). Indeed, Schumpeter's work is still considered the most important and creative theory of entrepreneurship that is currently in existence (Swedberg 2006).

Schumpeter was the first to recognize the important role that entrepreneurs play in the innovation process. To Schumpeter, being an entrepreneur involved doing something new and 'breaking the mould'. Innovation arose from 'new combinations' of resources (physical, intellectual, capital, human and social) and entrepreneurs were responsible for bringing these new combinations about. According to Schumpeter, there are normally too many new combinations of resources to analyse rationally so the talent of the entrepreneur is to intuitively recognize those which can create the greatest value and competitive advantage in the market. It is this intuitive capacity to see the value of new ideas that distinguishes a good and bad entrepreneur.

Schumpeter saw the process 'creative destruction' as central to the innovation process. This involves obsolete ideas, technologies, products, services and firms being swept aside and replaced by more up-to-date and superior successors. In explaining the process of creative destruction, Schumpeter argued that most of the time innovation occurs incrementally through people improving on existing products and services within a well-defined, stable and widely accepted set of social market norms. Some people manage these rules better than others but occasionally this framework is dislocated by a major development or event which changes the basic conditions of the market and redefines the 'rules of the game'. Research shows that these potential triggers for discontinuity are difficult to predict in advance and much research shows that organizations have in-built imperfections and barriers which prevent them from seeing these shifts in advance. Chapters 5 and 6 discuss these barriers in much more detail. The key point here is that organizations need to be continually scanning their environments to detect such events and, given that this is very difficult, need to have adaptive capacities to change their established organizational routines when the inevitable unexpected happens.

Schumpeter also argued that this process of creative destruction must come 'from within', an emphasis which drew an important distinction between the process of 'adaptation' (externally driven change forced upon the organization from outside) and 'development' (internally driven change generated by internal initiative). To Schumpeter 'adaptation' is not innovative (since an organization is effectively dragged along) whereas 'development' is innovative (because it is voluntary, creative, pro-active and breaks new ground). Introducing the idea of the entrepreneur into modern management vocabulary, Schumpeter saw this person as the 'man of action' who catalyses the innovation process. Challenging mainstream economics, which was (and still is) dominated by equilibrium models, Schumpeter argued that an entrepreneur's key occupation is to search for opportunities to upset market equilibriums and, by doing so, create unique competitive advantage and excess profits (at least until others enter the market by copying and bring it back to equilibrium).

According to Schumpeter, the sociological factor that prevents innovation from happening is 'static people' who react negatively when someone tries to do something new. People react negatively because they feel psychologically threatened and are taken out of their comfort zone to do something that is not familiar and is therefore harder to do. In contrast to 'static people' who resist innovation and who stop when they have achieved their immediate needs or goals, the entrepreneur has no equilibrium point at which the energy invested equals the benefits gained. They are continually searching for new ideas and ways to improve what they are doing. These differences are summed up in Table 2.2.

Since Schumpeter's early work the concept of entrepreneurship has attracted considerable attention and has been refined. For example, Baumol (2010) has distinguished between the 'innovative entrepreneur' (who comes up with new ideas and puts them into practice) and the 'replicative entrepreneur' (who can be anyone launching a new business venture regardless of whether it is new or not). The replicative entrepreneur creates wealth by gaining a larger slice of

Table 2.2 Characteristics of entrepreneurs and non-entrepreneurs according to Schumpeter

Entrepreneurs	*Non-entrepreneurs*
Dynamic – seeks change	Static – reacts negatively to change
Leverages disequilibrium	Seeks equilibrium
Does what is new	Repeats what has already been done
Active and energetic	Passive and low energy
Shows leadership	Follows
Creates new combinations of ideas	Accepts existing ways of doing things
Thrives on change	Feels uncomfortable with change
Battles resistance to new ideas	Battles new ideas
Intuitive and subjective	Rational and objective
Motivated by the joy of creation	Motivated by necessity
Insatiable appetite for new knowledge	Relies on past knowledge
Ambitious	Settles for mediocrity
Not bounded by existing resources, borrows resources to fund ideas	Works within existing resources, has no use for extra resources

(Adapted from Swedberg 2006: 29)

an existing pie (redistributive entrepreneurship) while the innovative entrepreneur creates wealth by making the pie bigger (productive entrepreneurship). According to Baumol, throughout history redistributive entrepreneurship, through wars and competition, has been by far the most common form of wealth accumulation. This is by definition a destructive activity undertaken through military type strategies which are designed to capture, hinder or even destroy a competitor's business. This, he argues, is a negative-sum game and history tells us that in the end everyone loses out by such activity. In contrast, productive entrepreneurship is a positive wealth creating activity which involves forging new markets or new space in existing markets by bringing new ideas and inventions to reality. Baumol argues that this is the type of entrepreneurship that Schumpeter was no doubt envisaging when he constructed his early theories.

More recent research into productive entrepreneurs by Muller and Becker (2012) identify eight attributes which are central to their success:

1. Motion
2. Preparation
3. Divergence
4. Commitment
5. Activation
6. Connection

7. Permeability
8. Attraction.

Motion

Entrepreneurs often break with routine and exhibit a high level of 'mobility' which puts them in the path of unfamiliar situations, allows them to mix with unfamiliar people, exposes them to the unknown and unpredictable and allows them to discover things they weren't necessarily looking for. Most businesses are organized to restrict mobility by requiring employees to behave in predictable, controlled, patterned, orderly, accountable and measurable ways.

Preparation

While innovation may seem to imply a certain degree of spontaneity, it also requires some level of preparation. Managers who are prepared for the unexpected to happen are better able to detect and chase unexpected leads, explore alternative paths and to see the importance of patterns which others might miss.

Divergence

Divergence requires a willingness to change course, to challenge path dependencies, to chase new ideas and to explore unchartered waters. Successful entrepreneurs relentlessly search for new ideas. While they don't expect every idea to work out, it is the variety of options that they explore that increases their chance of a discovery that may just lead to success.

Commitment

Commitment is a sense of dedication and purpose which enables successful entrepreneurs to discriminate between good and bad ideas and to overcome resistance to their implementation.

Activation

Most people live in monochromic time. They live highly scheduled, compartmentalized and pressured lives and are driven by the need to complete tasks according to strict priorities and planned deadlines. Since time is a precious resource, the need to meet targets and schedules take precedence over building relationships and a sharp distinction is drawn between work and personal lives. In contrast, entrepreneurs live by polychronic time. They prioritize personal relationships, are flexible with their commitments, make little distinction between work and personal lives and tackle multiple tasks simultaneously.

Connection

Professional networking sites like LinkedIn and social media like Facebook present almost limitless opportunities for people to connect. Entrepreneurs are masters at unlocking the value in these networks. By 'positioning' themselves centrally in these networks they maximize their access to information and can control its flow to their advantage.

Permeability

Entrepreneurs build permeable organizations which engage with an open and ongoing conversation with the outside world. They recognize that effective communication is a two-way rather than a one-way process and that 'everyone' should be engaged in this process – not just designated senior individuals or departments.

Attraction

Entrepreneurs have mastered the art of attraction – the ability to draw the most valuable people, ideas, information and opportunities to them rather than to their competitors. Entrepreneurs are like magnets for new ideas. They make themselves irresistible to other people by being clear about their purpose, values and beliefs and by portraying an infectious passion, enthusiasm and positive image. By doing this they create a 'reason' for others to connect to them and draw them to their cause.

Sustainable entrepreneurship

More recently building on Schumpeter's early work we have seen the development of a new branch of entrepreneurship called 'sustainable entrepreneurship'. The interest here is the application of an entrepreneurial approach towards meeting societal and ecological goals (Schaltegger and Wagner 2011). This new type of entrepreneurship has built on the early and narrower concepts of 'eco-entrepreneurship' (focused on ecological issues) and 'social entrepreneurship' (focused on social issues). Like all entrepreneurs, sustainability entrepreneurs seek to creatively destroy existing unsustainable products and services and consumption patterns and replace them with sustainable alternatives. Their key goal is to integrate ideas of sustainable development into the logic and strategy of businesses by demonstrating that firms can earn money through solving social and ecological problems. Loosemore and Phua's (2011) analysis of corporate social responsibility strategy in construction discusses this trend. They argue that while the major firms in the industry are making good progress in this area, the majority of the construction industry is still driven by the bottom-line and struggle to see the benefits that a more sustainable business model can bring to their business. There are many reasons for this discussed in their book. However, of relevance here is their finding that one of the many reasons why firms have struggled to incorporate corporate social responsibility into their business strategies is that many have relied too heavily on the activities of a dedicated corporate social responsibility manager. The problem with this approach is that corporate social responsibility then becomes an 'add-on' or a separate fiefdom which separates it from the mainstream activities of the business. In contrast, sustainability entrepreneurship theory would argue that to be successful, sustainability should not be separated off but should be imbedded into everyone's day-to-day functions. Only in this way can it constitute and shape the heart, soul and face of their company and imbue it with the spirit and process of creating products and services which bring about positive economic, social, cultural and ecological change. Sustainable entrepreneurs influence company policy and practices through their own personal goals, show personal mastery of the subject, demonstrate passion and unwavering commitment towards their cause and consider the gap between their vision and current reality as an opportunity rather than a problem. Sustainable entrepreneurship is challenging because in realizing this goal, sustainability entrepreneurs need to consider the demands of a wide range of stakeholders whose demands go far beyond the narrow economic interests of shareholders.

Contemporary models of innovation

Since Schumpeter's early ideas, many researchers have attempted to produce models to represent the innovation process. According to Dodgson et al. (2005), the major analytical approaches underpinning these models can be categorized according to six key criteria:

1. Sources of innovation (science driven supply-push v market driven demand-pull models)
2. Nature and extent of innovation (radical v incremental)
3. Type of innovation (product, service, process)
4. Systems of innovation (national, regional, sectoral, networks)
5. Innovation processes (R&D, innovation strategy, knowledge management)
6. Innovation outcomes (diffusion and learning).

Tidd (2006) identifies five generations of models in the innovation literature.

1. First generation models emerged after the Second World War out of the many large-scale scientific R&D projects in hi-tech industries like nuclear engineering, space exploration and medicine. They saw innovation as a 'supply–push' process where innovation was driven by technical capabilities to create new products not by the need for them in the marketplace. These models were highly linear and structured.
2. Second generation models emerged in the 1970s and saw innovation as a 'demand–pull' process. These models saw innovation being driven by

consumer demands for new products and services. They emerged when the impact of formal R&D on industrial innovations started to diminish and when market forces and growing consumerism took over as the main drivers of innovation in society.

3. Third generation 'coupling' models emerged in the 1980s to challenge the linearity of earlier models. These focused on the internal workings of organizations, particularly the effective feedback loops and communications needed around technological and market knowledge. These third generation models were influenced by the lessons in Japanese industry which highlighted the need for integration and teamwork between research, production, engineering and marketing functions, facilitated through techniques such as total quality management, just-in-time and lean production.
4. Fourth generation models emerged in the 1990s in response to advances in technologies and increasing knowledge about the types of organizational changes needed to exploit them. These 'parallel lines' models emphasized the need for integration of information flows between multiple sources of innovation in supply and demand chains, working with customers and business partners and the importance of linkages and alliances.
5. Fifth generation 'systems integration' models emphasize the role of systems integration, social capital and business networks in facilitating the co-creation of knowledge across different knowledge domains. These recent models argue that innovation occurs not so much from the ownership of assets but from dynamic capabilities in manipulating resources and working in integrated global networks of specialized firms.

More recently, Anthony (2012) has identified a sixth-generation of innovation which is based on company size and which is being driven by large multinationals like IBM, Unilever, Google, Microsoft and Apple. The idea that larger companies are now becoming the key drivers of innovation in society confounds the widespread belief that large companies are too bureaucratic, rigid and hierarchical to innovate. These beliefs originate from very early models of innovation which promoted the idea of the lone-inventor working alone to perfect an idea driven by a personal passion and insight. Innovators like Albert Einstein (who developed the Theory of Relativity), Hendry Ford (who developed the concept of mass production), the Wright brothers (who developed the first airplane) and Alexander Fleming (who discovered penicillin) come to mind when one thinks of this approach. However, Antony argues that as mass production grew, the complexity and increasing costs of innovation forced it out of the capabilities of individuals working alone and into the hands of large corporations which led a second era of innovation driven by large research and development laboratories in companies like DuPont, General Electric, Proctor and Gamble and 3M. As these companies grew and became overly bureaucratic during the 1960s and 1970s, innovators began to leave and form their own small-to-medium-sized enterprises funded by a burgeoning venture capital industry. This led to the formation of companies like Microsoft, Apple, Amazon, Starbucks and Cisco which themselves have matured leading us into a new era of innovation in which entrepreneurial individuals in large companies are using their employer's enormous resources, scale and agility to develop solutions to truly global challenges such as sustainable development and climate change. Household names like Steve Jobs, Bill Gates and Richard Branson characterize this new era of 'innovation superstars' who have the vision, global outreach and resources to develop and implement global innovations such as IBM's Smarter Cities initiative which has helped entire cities more efficiently manage energy, water, traffic, poverty and crime. IBM claims that in Stockholm, the bundling of its advanced technologies with related transport services has reduced carbon emissions by 17 per cent and traffic delays by 50 per cent. Rather than building a $1 billion tunnel to alleviate traffic congestion and disrupt the city for years, IBM's Smart Cities team constructed an operations centre which use sensors to monitor every car and offer incentives for drivers to avoid congested routes or used public transport. The Smarter Cities initiative was born in 2006 out of its 'InnovationJam' online brainstorming sessions which involved 150,000 IBM employees, family members, business partners and clients. IBM established this initiative as a way of giving its network of employees, who were becoming increasingly remote from their offices, a sense of participation in their business. Running globally for 72 hours at a time, and off a network of interlinked bulletin boards and related Web pages on IBM's intranet, the intent of these exercises is to seek substantive answers to important questions and to provide insights into future business growth areas for IBM to invest in. IBM invested $100 million in ideas that emerged out of the jam in areas like water, energy, transportation and healthcare. Originally called 'Instrumental Planet', it was a serendipitous visit to Masdar City in Abu Dhabi, an urban planning project which sought to create a carbon neutral and zero waste environment, which led to the idea of Smarter Cities, a centralized integrated operations centre for a city.

Regardless of the model one subscribes to, contemporary innovation models move us beyond the simple dualistic problem of whether innovation is supply-pushed (based on new technological possibilities) or demand-pulled (based on social needs and market requirements). Through empirical research we now know that innovation is not an individual act and does not just happen within the industrial supply-side, or as a result of the articulation of user demand, but through a complex set of processes which link many different actors together. We also know that the most successful innovations occur at the boundaries of organizations and often in serendipitous unplanned ways. This recent work challenges early ideas which conceived innovation as a deterministic process. Instead, innovation is seen as a socially constructed phenomenon arising out of many conscious and unconscious 'choices'. In other words, innovation does not necessarily emerge by following a predetermined logic but through many possible paths, each having differing implications for society and for particular social groups. As Wenda et al. (2005) points out, there is a 'mutual shaping' of new ideas by firms and society.

In the broadest sense, this evolution of innovation thinking reflects a move from closed-systems to open-systems thinking. The early closed-system paradigm saw innovation as a closely controlled process where organizations 'go it alone', seeking to control the generation of their own ideas, as well as their production, marketing, distribution, servicing, financing and support. Typically, this model involved the creation of a dedicated research and development (R&D) department with the responsibility to control the whole new product development cycle inside the company. In contrast, contemporary open-system paradigms recognize that new knowledge is no longer proprietary to one firm but resides in employees, suppliers, customers, competitors and universities etc. Contemporary theories of open innovation, user innovation, cumulative innovation and distributed innovation all argue that innovation occurs in collaboration with partners from outside an organization and that firms should use both internal and external networks and resources to develop, implement and bring new ideas to market. Today, the boundaries between a firm and its environment are seen as highly permeable, enabling the transfer of knowledge between the firm and a world of widely distributed knowledge. The implication of these ideas is that companies cannot afford to rely entirely on their own internal resources, but should instead buy or license inventions from other companies as part of their own innovation process. Similarly, internal inventions should be taken outside the company (e.g. through licensing, joint ventures and spin-offs etc.). As Tom Peters and Robert Waterman pointed out in their best-selling book *In Search of Excellence*, 'Innovation comes only from readily and seamlessly sharing information rather than hoarding it' (Peters and Waterman 1982:164).

The psychology of business innovation

Tom Peters is a neat segue into the modern drivers of innovation in business. Tom Peters, like many modern-day management gurus, has benefited from what Panati (1991) argues is one of the most defining characteristics of the late twentieth century – the emergence of brands and fashions as determinants of purchasing behaviour. As Stewart's (2009) critique of the management consulting industry shows, the concept of brands and fashions has not escaped the business community and has been responsible for the dramatic growth in the global management consulting industry over the last two decades. Business trends such as downsizing, business process re-engineering, supply chain management, total quality management and lean manufacturing have all seen their day in the sun based on claims to produce miraculous results. For example, Heller (1993: 40) argued that 'there is nothing that TQM's tools and techniques are not able to tackle' and reported one typical company where service had improved by 30 per cent and inventories were down by 80 per cent. Oliver's (1993) claims of benefits from TQM were even more incredible, citing 40–90 per cent improvements in process turn-around times, 10 per cent reductions in unit costs, 50–80 per cent quality improvements, higher staff and customer satisfaction and increased business. And Richardson (1996) claimed that by re-engineering their production processes, one company was able to reduce its lead times from 105 to 32 days, double inventory turn-around and increase productivity by 50 per cent. The construction sector has not been immune from these trends. For example, Marosszeky (1999: 343) asserted that 'For the construction industry to be truly successful, it needs to emulate this manufacturing culture, adopting similar methods and goals, as they are as relevant to construction as to any other industry'.

Despite the claims that imported management techniques could produce huge improvements in construction efficiency and performance, it was often the case that these ideas were based on nothing more than anecdote and an absence of any empirical evidence to verify any measurable and enduring impact in practice. For example, a study of 3500 're-engineered' companies in Australia revealed that 60 per cent failed to see any productivity improvements (Savery and Lucks, 2000). Indeed, there is significant evidence that many of these so-called innovations can have the opposite effect. In the construction sector, Green (1998, 1999, 2006, 2009, 2011) and Ness (2010) have been particularly critical of the uncritical adoption of these management 'recipes' from manufacturing, producing evidence that they reinforce 'command-and-control' cultures and are counterproductive for people who work in the industry.

Given the tendency of many managers in the construction sector to ignore this evidence and to willingly spend large amounts of money and time to embrace these ideas without any rigorous analysis, we explore below the reasons why people are so seduced by their claims.

The supply for new ideas

Huczynski (1993) explains that there is both a supply and demand for new ideas in business. The supply side has been driven by the rapid growth of the management consulting industry in recent years which has created a lucrative market for new management ideas. As Kipping and Engwall (2002) note, in 1980 there were only five consulting firms which employed little more than 1,000 management consultants worldwide. By the 1990s there were more than thirty firms of this size employing many hundreds of thousands of aspiring management consultants. Huczynski (1993) showed that managerial ideas, like any product, have a limited shelf-life. So the growth of the management consulting industry requires them to generate a constant supply of new ideas for managers to consume. Since research shows that most managers only read narrowly for an average of four hours a week, the attractiveness of these ideas depends upon them being packaged in a simple and convenient way. It is even better if an idea offers a unified set of principles which can be applied to a wide range of situations. Shrewd management consultants know that most managers reduce, as far as possible, the information needed to make a decision, welcoming any relief from lengthy searches for solutions to their problems. And the burden of decision-making is reduced even further if a manager can follow the lead of other managers, which is why decision-makers tend to give disproportionate weighting to endorsements and anecdotal information about successes and failures in other companies than to statistical data which involves lengthy analysis. So it is no coincidence that new management ideas are often accompanied by mythical stories of staggering success in the world's 'best practice' organizations, rather than by solid data. After all, this is what the market wants.

The demand for new ideas

From a demand perspective, Huczynski (1993) found several reasons why managers succumb to the temptations of the latest management fad.

1. The adoption of a new idea is likely to be a career enhancer, increasing the visibility of the person championing it within an organization.
2. A new idea from a famous management guru can be used as a 'managerial defence' since no manager can be reasonably blamed for adopting the latest management fad, even if it does not work.
3. The idea might be seen as offering quick results to complex problems. Because many managers are drawn from technical specialities such as engineering, law and finance and have little managerial know-how, they are particularly attracted to quick-fix solutions which can defend them against accusations of ignorance.
4. The idea may satisfy the increasing need to be associated with 'best practice' which is why the new ideas are often sold first to prominent companies such as ICI or IBM. Smaller organizations will then follow suit, wanting to be associated with them. All fads depend upon the phenomenon of crowd behaviour.

While the above factors relate to individual managers and their personal need for new ideas, demand for new ideas is also driven by an array of new organizational imperatives. Increasing global competition, declining profitability and continual demands for business growth and increased productivity represent powerful reasons to invest in the types of radical change promised by many new management techniques. In such an intensely competitive business environment, ignoring any potential source of new ideas could mean giving competitors an advantage. Today's managers are under unprecedented pressure to prove to increasingly informed and empowered customers and shareholders that they are at least doing something constructive to drive up performance. And to portray themselves as innovative and pioneering, it makes every sense to adopt what appears to be the latest management idea, even if the focus becomes the technique itself, irrespective of what value it can add to an organization.

The fear of not changing

During the 1970s and 1980s influential texts such as Toffler's (1970) *Future Shock*, Kanter's (1983) *The Change Masters*, Peters and Waterman's (1982) *In*

Search of Excellence, and Peters' (1989) *Thriving on Chaos* convinced managers that their lives would never be the same. In the future, all businesses would be faced with unprecedented, accelerating and uncontrollable technological, economic and political change. The dominant management rhetoric during this time argued that the path to prosperity rested on 'courageous' managers challenging the status quo and embarking upon a constant process of revolution, self-examination, renewal, innovation and radical change. Peters and Waterman's *In Search of Excellence* sold over five million copies and warned that the world's largest and most successful companies were most vulnerable to this new environment. This message was reinforced by writers like Pascale (1991: 11) whose catch-phrase '*nothing fails like success*' became part of any self-respecting manager's vocabulary.

The brazen warnings struck fear into the heart of the business community and produced a new and lucrative market for managerial 'wonder drugs' to help managers successfully bring about organizational change. The business community was bombarded with a proliferation of new management buzzwords such as: management by objectives; decentralization; delayering; supply chain management; value chains analysis; down-sizing; diversification; restructuring; and quality circles. And in the enthusiasm for change, these ideas were seldom challenged and were embraced by the majority of managers with uncritical enthusiasm and zeal. This was despite accumulating evidence that cast doubt on the veracity of the revolution mentality. For example, Collins and Porras's (1994) analysis of eighteen successful companies such as 3M, American Express, Motorola and Proctor and Gamble, which had consistently outperformed the American Stock Market over a number of decades, found that change was never an over-riding factor in how these companies operated. Instead, the one factor that linked these companies was their zealous preservation of their core ideology and identity which in some cases had remained unchanged for over one hundred years. Importantly these core values emphasized people rather than processes, which was in stark contrast to many of the management fads which developed during that time.

The need for simplicity

A common characteristic of most management fads is that they reduce management to a few simple profound truths and universal laws. This is not a coincidence because as Watson (1994) showed, most managers cannot find the time to think through in detail every problem they face. Consequently they tend to act intuitively and without reflection, repeating or adapting decisions which have served them well in the past. Furthermore, in an attempt to impose some order upon an increasingly chaotic world, most managers tend to turn to the predictability and safety of the simple universal solutions which many management buzzwords promise.

However, while the process of simplification is a necessary and understandable defence against increasing business complexity, it is also potentially dangerous. As Watson (1994) showed, simplicity often occurs at the expense of getting people together, sharing ideas and developing solutions to shared problems. In other words, managers need to see the wood as well as the trees so that they can understand the overall context that they operate within.

Evangelism

One reason for the zealous uncritical enthusiasm which managers show for the latest management fad can be found in the highly evangelical, rhetorical and one-sided language that is used to promote them. Extreme and unfounded claims of successes are often made around these new ideas with few, if any, failures ever being mentioned. Hilmer and Donaldson (1996) warn that such dogma fosters superficiality, putting the reasoning powers of otherwise skeptical, thoughtful and pragmatic managers on hold. This causes managers to turn to management gurus in the same way that the ancient Greeks turned to their gods. For example, during the 1980s many managers would openly boast that they were a disciple of the 'Demming gospel', following his fourteen simple steps in the same way that Christians followed Moses' Ten Commandments. In the construction industry similar zeal has accompanied the introduction of many new management ideas. For example, during the 1990s 'partnering' gained much attention attracting a huge following of believers who promoted the idea as the panacea to all the industry's woes. Partnering was a new collaborative approach to procurement imported from manufacturing and its widespread popularity came close on the heels of the highly influential Latham Report *Constructing the Team* (Latham 1994) which criticized the industry for its confrontation approach to building procurement. However, as Bresnen and Marshall's (2000) critical appraisal of partnering in the construction sector revealed, partnering was not the universal panacea it was portrayed to be. Many partnering projects were simply unsustainable and made up of A-teams which could not be replicated elsewhere. To be successful, the implementation of partnering should be carefully thought through and supported by a clear sense of strategic direction and, most importantly, senior management commitment and resources. Without this, it could be positively counter-productive to introduce an open trusting approach on an industry that had been steeped in confrontation and conflict for so long. Furthermore, successful partnering requires an understanding of the likely impact upon individual and group behaviours, motivations and interests and a full appreciation of the long-term and complex dynamics of the implementation processes.

Innovation is 'messy'

As we discussed above, early models of innovation saw it as a staged, highly linear and sequential process. Dodgson and Gann (2010) draw an analogy to a production line or relay race where one part of the organization starts the innovation process by creating an idea (typically the R&D department) and hands it to another (such as design) who then hands it to another (production) and finally to commercialization (marketing and sales). Barrett et al. (2008) argue that this rational and linear view of innovation typifies the construction innovation literature. However, as discussed above, recent research indicates that the innovation process is far from linear but is messy, uncontrollable, unpredictable and full of feedback loops and setbacks. This view is endorsed by Fuglsang (2008) who points out that it is often difficult to define the beginning and end of the innovation process and that innovation is highly interactive and cyclical, involving many people with multiple interests 'co-creating' new knowledge under the influence of market and many other social, political and organizational forces. Dodgson and Gann (2010) use the analogy of a rugby game where different players with different attributes, characteristics and skills simultaneously perform different roles but collectively aim towards one common objective.

Innovation therefore involves both intrapreneurship (working with internal stakeholders to seek opportunities and implement new ideas into existing systems) and entrepreneurship (seeing and seizing external opportunities for change). Through this collaborative process, a new idea develops with the input of many peoples' thoughts and opinions into several ideas which are then further developed and aggregated gradually through analysis, testing and experimentation into something solid. Shocks and surprises are normal and progress can often be two steps forwards and one backwards. The most promising of these ideas are eventually selected and scaled up through further testing and development (normally through market mechanisms, social and business networks). At the same time, champions for each idea are garnering support both political and resources and managing multiple pressures from peers, staff and managers to move in different directions. This is essential to assist in the creation of a business case and to prompt progress towards the final diffusion stage where the idea is communicated and implemented into a social system to bring about change.

Innovation diffusion

Widen et al. (2008) and Brandon and Shu-Ling (2008) argue that the construction literature on innovation tends to focus on the early creation and development phases of the innovation process and ignores the way in which new ideas are 'diffused' into the wider community. Diffusion is the process by which an innovation's acceptance spreads through a social system bringing about a change in its structure and functioning for the benefits of its stakeholders. This is often referred to as the 'commercialization phase' of the innovation process, where ideas and inventions are moved out of laboratories, design and production into the marketplace. While the outcomes of innovation are not always commercial but also social, this is often the most difficult stage of the innovation process. As Howard Aiken, the original conceptual designer behind IBM's first computer, once said, 'Don't worry about people stealing your ideas. If your ideas are any good, you'll have to ram them down people's throats.'

The importance of thinking about diffusion is demonstrated by the cautionary tale of how Xerox missed one of the biggest opportunities in modern business history, when it failed to successfully communicate the value of the world's first personal computer in 1973. Few people realize that this was created by a team of Xerox researchers in its Palo Alto Research Centre in California and that it included a mouse, an excellent user interface, a laser printer and an ethernet. As Morris (2008) points out, Xerox failed to communicate the significance of their monumental achievement and lost out to IBM, Microsoft and Apple who most people now assume to be the pioneers of PC computers.

A key characteristic of diffusion is that it is a process that occurs within a social system. Rogers' (1962) synthesis of research from over 508 diffusion studies produced the first theory of innovation diffusion which is still relevant today. He found that the diffusion process follows an s-curve and that with successive groups of consumers adopting the new technology (shown by the bell curve), its market share (the s-curve) will eventually reach saturation level (see Figure 2.1). According to Rogers, an innovation starts slowly with a few adopters then increases at a faster rate until most have adopted the idea then slows down again. In reality, the majority of new ideas are of course never adopted in mass and an important factor explaining the slow start of the diffusion curve is the low relative advantage of a new idea when it is first introduced.

First-movers typically represent around 2.5 per cent of the business population to eventually adopt an innovation. These firms are seen as innovators, mavericks and risk takers in their industry, highly imaginative, creative,

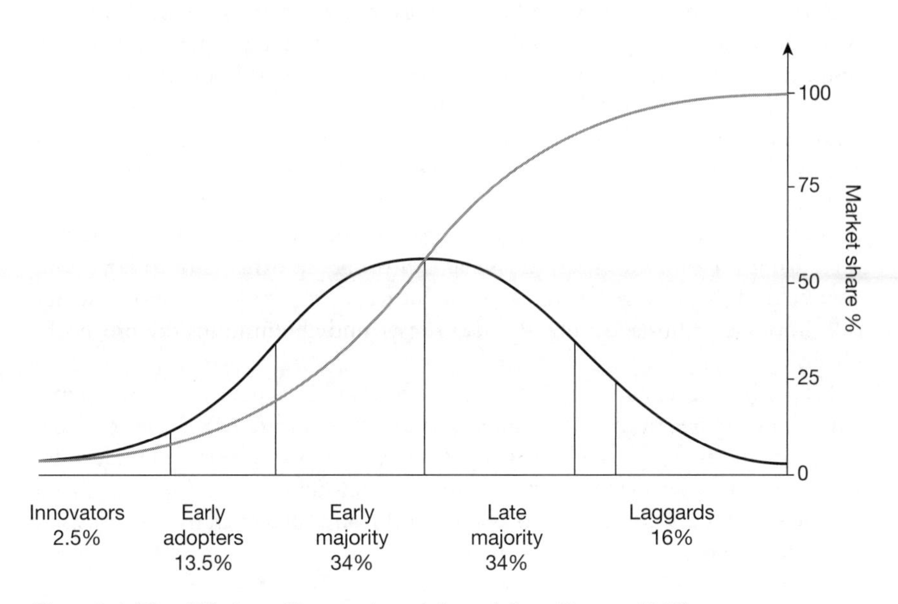

Figure 2.1 The diffusion of innovations (adapted from Rogers 1962)

venturesome, youthful, cosmopolitan, open, expansive and widely connected. Outside of construction, one would think of firms such as Xerox, Apple, Microsoft, IBM and Google. These firms are highly sensitive and responsive to new ideas from a variety of knowledge domains, actively lobby for change and regularly become aware of new ideas before their competitors. Financial security and flexibility also enable these firms to be more decisive in adopting new ideas and in accepting and adjusting to change.

Gann (2003) argues that most firms in the construction sector tend to be adopters of innovations from other industries rather than the original sources of innovation. Early adopters typically represent about 13.5 per cent of the population and are a little more cautious than first movers, although well ahead of the majority in taking up a new idea. They tend to be more securely integrated into their social system than first movers and command a high degree of respect by their industry peers and customers since they embody the careful, objective, selective and successful use of new ideas. They are also seen as less erratic and more reliable indicators of trends than first movers and tend to act as an industry reference point. Since they are seen as more cautious and less easily swayed by the latest fads than first movers, they also tend to act as industry legitimators and opinion leaders, having a higher degree of influence over the direction of their industry. Finally, while early adopters may not experience the potential super profits of first movers, they tend to have a higher rate of success and their innovations also tend to have a higher degree of credibility and resilience.

The early majority which typically represent about 34 per cent of the population of eventual adopters are very deliberate in their decisions to adopt a new idea. They are not known as leaders in their industry and reflect rather than drive industry trends. They tend to carefully study and copy the majority only when ideas are well proven and they tend to deliver consistent and reliable results using well-proven but contemporary methods with little risk.

The late majority or followers typically represent 34 per cent of the population and are inward looking. They only move when there is no other option and when the weight of opinion to change is irresistible. They fear change and have a very low risk appetite and when they do eventually move they are ponderous and tend to regard with skepticism the value of any change, measuring its value in purely financial terms. These firms tend to lag a long way behind the average firm in implementing any new ideas or processes and only adopt after first ignoring, discrediting or finding numerous reasons not to change.

Laggards, resisters and rejecters represent around 16 per cent of the population and are the last to do anything. They are very traditional in their outlook, backward in most ways, and skeptical of change. They resist or even undermine new ideas and are openly hostile to anyone wishing to bring it about. These are the social drop-outs of the corporate world, alienated from the majority and not taken seriously.

Diffusion models

Fundamentally, the diffusion process is about communication and traditional models of diffusion have relied heavily on the centre-periphery model. This conceives the innovation diffusion process originating at the centre of a network which communicates the idea to other actors at the periphery of the network (much like the spokes in a wheel with a central hub as illustrated in Figure 2.2).

Clearly, the effectiveness of this centre-periphery system depends on the energy, enthusiasm, commitment and resources of the person at the centre of the wheel. It also depends on the number of points at the periphery, the length of the spokes through which diffusion takes place and on the effectiveness of communication along those paths. While this approach to innovation diffusion appears to be fast and efficient, it starts to break down as business environments become more complex. So when operating in complex environments, organizations tend to move to a franchise model which involves the proliferation of centres (see Figure 2.3)

In this franchise model of innovation diffusion each peripheral node represents a separate distributing network. In essence, the system retains the centre-periphery structure but differentiates between primary and secondary centres which have distinct roles in the innovation diffusion process. The role of secondary centres is to engage in the diffusion of innovations while primary centres select territories for expansion and methodologies for diffusion, train

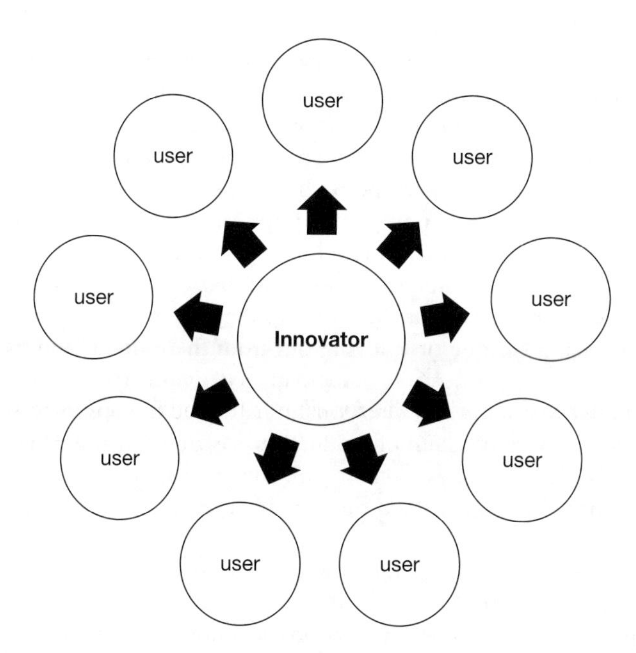

Figure 2.2 The centre-periphery model of innovation diffusion (adapted from Schon 1983: 192)

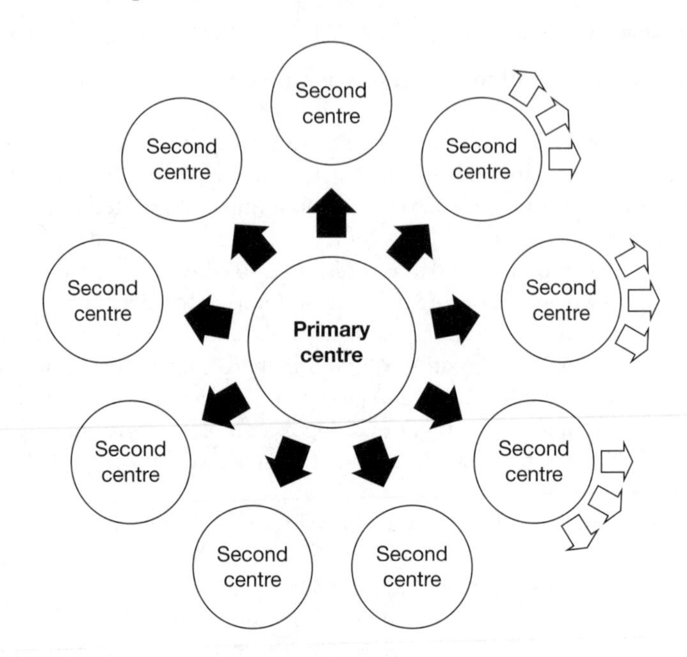

Figure 2.3 Franchise model of innovation diffusion (adapted from Schon 1983: 193)

and incubate the new agents and champions of diffusion and support and resource the new outposts. The primary centres also coordinate and manage communications between the secondary centres and monitor and manage their performance. In effect, this organization becomes a constellation of semi-independent centres which acts to increase the reach and efficiency of the diffusion system as a whole. This approach has been used by colonizers throughout history to diffuse their power and political or religious ideologies. For example, the Romans used this model to convert new territories to Roman rule and Christian and other religious missionaries have used it to convert distant peoples to their religion. The British Empire was also built on this model with explorers and conquerors moving out from the mother land from which they were monitored, resourced and managed. More recently, in business, companies like Coca Cola and MacDonalds have used this approach to replicate a single tried-and-tested system of production, distribution and marketing. The simplicity and uniformity of the central message and approach permits a system of deployment which has enabled these companies to quickly dominate world markets.

The effectiveness of franchising, which has become the dominant model of diffusion in a globalized business world, rests on the management abilities of the primary centre to support the secondary centres. Success also rests on the alignment between their individual cultures, processes and systems. Research shows that failure in these systems tends to takes the form of intra-firm conflict, fragmentation, insubordination and classic problems of centre-branch conflict which lead to eventual disintegration into independent centres. However, research also indicates that ironically, this may be an inevitable and necessary process to enable peripheral centres to integrate into local markets and ultimately enable the centre to survive. The trick in managing these dispersed types of networked organizations then becomes balancing the need for centralized control and integration and peripheral differentiation.

The speed of diffusion

Many researchers have explored the factors that determine the speed of innovation diffusion through innovation networks. For example, in the construction sector, Gambatese and Hallowell (2011) found that technological innovations diffused more readily when the amount of training required to understand them and the perceived risk of failure was low. This was also the case when the perceived benefits of innovations were clearly pointed out (particularly, cost savings, quality improvement and increased productivity). Other studies have found the timing of the innovation to be important to effective diffusion as well as other factors such as the existence of supporting innovations, the complexity of the innovation, the trialability of the innovation, and the amount of supervision and coordination needed to support it.

Stages in the diffusion process

The diffusion of an innovation doesn't happen in one step and to help us understand the process in more detail, Rogers (1995) proposes a model which comprises of the following five stages.

1. Knowledge stage (occurs when an individual is exposed to an innovation's existence and gains some understanding of how it works)
2. Persuasion stage (occurs when an individual forms a favourable or unfavourable perception or attitude towards the innovation)
3. Decision-making stage (when they decide to adopt or reject the innovation)
4. Implementation stage (when an individual puts the innovation into use)
5. Confirmation stage (when an individual seeks reinforcement of the decision to adopt and continues with or rejects or reverses that decision).

The knowledge stage

The relationship between knowledge and innovation is not as straightforward as it may seem. Sometimes people must have a need for the innovation before they become aware of it (demand-led) but for others the innovation itself can create a need that was not previously there (supply–push). We also know that people tend to selectively expose themselves to certain types of information that tend to confirm their existing beliefs. Festinger (1957) developed the theory of cognitive dissonance to explain this process and his theory indicates that people tend to avoid information which makes them feel uncomfortable (selective perception). For this reason, it is not necessarily those channels that reach the widest audiences which are the best diffusers of innovation. More often than not, the best channels are those that facilitate face-to-face communication. So in the diffusion process there is an inherent tension between the need to reach wide audiences and the need to persuade people to adopt. For example, while mass media may be effective at communicating with large numbers of people, it is not necessarily good at persuading them to adopt a new idea.

In providing knowledge to persuade people to adopt, we also know that some social networks are better at diffusing ideas than others. In helping us understand this process, research in the area of social network analysis can be very helpful (Freeman 1979; Rogers and Kincaid 1981; Freeman 2010). This research has shown conclusively that dense networks (where everyone is connected to everyone else directly) facilitate the most rapid transfer of ideas between the greatest number of people. However, we also know that it is hard to control the flow of ideas in such networks and that they are not the most efficient systems because there is a considerable amount of repetition and wastage in the transfer of ideas. These problems can be avoided by highly centralized networks (where communications either flow to or emanate from one person) but this depends heavily on the capabilities of the person at the centre. In effect, centrality affords an innovator the power to influence others in a network, communicate directly with more people and act as opinion leaders or change agents. Finally, we know there are people who act as valves in social networks. They sit 'between' groups of people and have the power to control information flows between them. So by understanding and influencing the people who occupy such positions of power, innovators can greatly facilitate the diffusion of their new ideas into the social systems they are seeking to influence.

The final property of social networks which influences the diffusion process is heterphily and homophily. First defined over a century ago by Gabriel Tarde (1890), homophily is the degree to which a pair of people in a network are similar in various attributes such as a beliefs, education, age, background, social status etc. Heterophily is the opposite. Tarde noticed that social relations are closer and communications are much more effective between people who are similar than those who are different. So logically one would expect homophilious networks to aid the diffusion process. However, research has found that in many organizations homophily can be a barrier to diffusion because organizations tend to be structured with similar people operating at the same horizontal levels. The key to overcoming this is to ensure homophily exist between, as well as within, organizational levels. Having said this there is inevitably a trade-off between heterphily and homophily in designing organizations. As Granovetter's (1973) *Theory of Weak Ties* tells us, most innovative ideas do not come from those around us (that we are closest and most similar to), but from those who are distant and different from us. So while homophily aids the diffusion process, heterophily aids the creative process that lies at the source of innovation. Both are needed in innovative organizations.

The persuasion stage

Persuading people of the need for an innovation can be difficult. Most people do not evaluate an innovation rationally on the basis of objective scientific studies but on subjective evaluations conveyed by other people (especially those who have already adopted the innovation). Whereas the main mental activity during the knowledge stage is cognitive (or knowing) the dominant mental activity during the persuasion stage is affective (feeling). At this stage, the person becomes more psychologically connected with the innovation and actively seeks more 'evaluative' information about it. This allows them to form a positive or negative attitude towards a new idea and to make a decision about adopting it or not. It is widely recognized that different people need different amounts of time to get through this stage. As we pointed out above, early adopters are persuaded more easily than followers to get involved in an innovative idea. Simon Sinek (2009), best known for popularizing a concept of 'The Golden Circle', argues that persuading potential customers to adopt your new ideas is a complex process and that it takes considerable effort to get to the 'tipping point' where customers will suddenly fall into line and start adopting an innovation in mass. Using concepts from the biology of human decision-making to explain why we are inspired by some people, leaders, messages and organizations over others, Sinek argues that people who reach this point the fastest are those who realize that customers don't buy 'what' you do but 'why' you do it (the sizzle – not the sausage). This is because the 'why' message activates the centre of the brain (the Limbic brain) which is where emotions and loyalties are formed. Sinek argues that successful entrepreneurs are those who target a 'why' message to people with similar values and beliefs, thereby maximizing their chances of success in selling their idea.

Social reinforcement is also an important part of the persuasion process. Social learning theory tells us people make decisions, not only on the basis of their own experiences, but also on the experiences of others with whom they are socially connected (Bandura 1977; Lundvall 2010). Social interactions not only help to confirm one's own experiences but they also reduce the inherent uncertainty and risk of adopting a new idea. So it is at the persuasion stage of the innovation process that 'change agents' play their greatest role.

Change agents are typically independent (often external) people whose expertise, credibility and personal attributes assist in bringing about change in a social system. They play an important role in overcoming system 'norms' which often act as a barrier to innovation. Norms are a psychological concept which refer to the established accepted patterns of behaviour of a social system that determine what is acceptable to the system or not. If an innovation falls outside these norms, then it is unlikely to be accepted.

Change agents bring about change by influencing 'opinion leaders' in a social system. An opinion leader is someone who (consciously or unconsciously) influences other people's attitudes, norms and behaviours, not necessarily as a function of formal position or authority but more likely through social status, social connectivity, respect, competence, social accessibility and trust. Opinion leaders therefore exemplify and express a system's norms and are normally those people who have been deeply integrated within it for long period of time. In effect, opinion leaders shape other people's perceptions and determine the shape of the innovation adoption s-curve. The s-shape of the diffusion curve occurs because once opinion-leaders adopt and tell others about an innovation, the number of adopters tends to accelerate. Once a critical mass occurs there is then a tipping-point at which there are enough people who have adopted the innovation to ensure it is self-sustaining.

In simple terms the role of opinion leaders is to bring about change in a social system and Kurt Lewin's (1947) well-known research showed us that this involves three main stages:

1. Unfreezing
2. Changing
3. Freezing.

In the 'unfreezing' stage an individual becomes aware of a need to change. Inertia must be overcome and existing mindsets undone. In the 'changing' stage, the situation is diagnosed and new models of behaviour are explored, tested and implemented. This is typically a period of confusion and transition. People are aware that the old ways are being challenged but do not have a clear picture as to what we are replacing them with yet. Finally, the 'freezing' stage involves imbedding the new modes of behaviour into the social system, crystalizing new mindsets and monitoring them to ensure they stay changed.

While widely used, Lewin's three-step model of change has also attracted some criticism. For example, Burnes (2004) has argued that his work assumed that organizations operate in a stable state and that there is rarely time for the 'refreezing' stage in today's dynamic world where the process of organizational change is almost continuous. Instead, we should think about this final stage as being more flexible so that unfreezing for the next inevitable change might be easier.

Since Lewin's early work, research has also identified wide range of factors that prevent organizations changing (Stickland 1998). These factors include:

- distrust of change agents
- fear of change and of the unknown
- desire to maintain existing power structures
- complacency
- lack of resources to support change
- conflicts between individual and organizational goals.

Conversely it is widely accepted that change is more accepted when:

- it is understood
- when it does not threaten security
- when those affected have been involved in bringing it about
- when it follows other successful changes
- when it reduces work
- when the outcome is certain
- when implementation has been mutually planned
- when top management commitment is evident and strong.

We also know that there are certain organizational characteristics which are relevant to change. For example, older, larger, formalized, centralized and hierarchical organizational structures tend to resist change due to established ways of working (path dependencies). And we also know that certain types of people are more open to change than others. In particular, research shows that younger and more educated people in lower organizational positions tend to be more open to change than less educated, older and more senior people. Finally, as Baron and Greenberg (1992) argue, this resistance to change can come in many forms. For example, people may leave an organization, they may actively undermine, resist and encourage others to resist change, they may reluctantly accept change while not agreeing with it or they may enthusiastically accept it and actively encourage it. In the field of organizational development, numerous strategies have been developed to overcome such resistance. For example, Furnhman and Gunter (1994) propose a combination of seven strategies for dealing with resistance:

1. A 'fellowship strategy' (which involves extensive and time consuming discussions and events to discuss change)
2. A 'political struggle strategy' (which targets the power structures and those who have both informal and formal legitimacy to oppose or support change)
3. An 'economic strategy' (which assumes people are motivated by financial inducements to change)
4. An 'academic strategy' (which assumes people are persuaded by rigorous reports, facts, analysis, arguments and information)
5. An 'engineering strategy' (which assumes that people will be forced to change by changing the nature of their jobs)
6. A 'military strategy' (which is reliant on brute force)
7. A 'confrontational strategy' (which assumes that if you can mobilize anger in people to confront a change then they will bring it about).

The decision stage

The decision stage involves a person making a decision about whether to adopt or reject a new idea. This stage is often the most difficult in the innovation process and getting there can take a frustrating amount of time. As Rogers (1995) points out, most innovations diffuse at a disappointingly slow pace. While some innovations spread rapidly, others take many years to reach widespread adoption. Research by Spence (1994), Rogers (1995) and Baumol (2010) indicates that the time-lag between persuasion and adoption depends on a wide range of factors as described in Table 2.3.

Implementation

So far the innovation diffusion process has been only a mental exercise. In contrast, implementation is *the process that turns strategies and plans into actions in order to accomplish strategic objectives and goals.* For example, implementation may mean training people to use a new technology, manufacturing a new product or delivering a new service. Implementation is said to be the Achilles heel of strategy and is generally undertaken poorly. For this reason it has been the subject of considerable research. This research indicates that effective implementation starts with an effective strategic plan which considers and describes:

- what needs to achieved and why it needs to be achieved
- how committed leaders are to implementing it
- how it will be communicated
- who are the critical stakeholders who will implement it
- whether these stakeholders have sufficient buy-in to drive the plan forward
- how these stakeholders will be motivated to implement it
- whether appropriate resources are available to implement it
- the internal processes that are key to driving the plan forward
- the potential roadblocks to implementing and supporting the plan.

The most common reasons strategic plans fail include:

- poor planning
- unrealistic goals and deadlines
- under-resourcing
- poor risk management
- a lack of buy-in and ownership by stakeholders
- poor communication
- incompatibility with existing systems, structures and cultures
- other activities take priority
- poor monitoring
- no accountability
- a lack of empowerment.

There are a multitude of voluminous books on how to develop a successful business strategy. But in summary, to avoid these potential pitfalls it is wise to develop a strategic plan with input from key stakeholders. It is also important

Table 2.3 Diffusion time-lag determinants

Factors	*Effect*
Relative advantage	If a new idea is perceived as better than the idea it replaces it will be accepted faster
Compatibility	If a new idea is consistent with existing values and needs of users it will be accepted faster
Complexity	If a new idea is perceived as easy to use and understand it will be accepted faster
Trialability	If a new idea can be trialed without commitment it will be accepted faster
Observability	If the results of implementing a new idea are observable to others it will be accepted faster
Social structure of the adopting system	If the system into which a new idea is introduced is densely connected and highly centralized it will be accepted faster
Cost	If a new idea has a low cost of adoption it will be accepted faster
Complexity	Ideas that are easy to understand will be accepted faster
Visibility	If a new idea is open to inspection it will be accepted faster
Divisibility	If a new idea can be tried before adoption it will be accepted faster
Comparability	If past experiences with similar ideas are positive it will be accepted faster
Utility	Ideas with greater utility will be accepted faster
Dependency on collective action	Ideas that do not require coordinated decision-making and consensus for adoption will be accepted faster
Tangibility	Manufactured artifacts tend to be accepted and adopted faster than abstract ideas and services
Incrementality	Ideas that bring about incremental rather than radical change in existing routines will be accepted faster
Respect	Ideas that do not threaten or discredit earlier products, ideas or services will be accepted faster
Ownership protection	Ideas not protected by IP and copyrights rights will be accepted faster
Exclusivity	Ideas which are offered exclusively and which can provide competitive advantage will be accepted faster
Competition	Ideas that come from non-competitors are likely to be accepted faster
Dislocation	Ideas that do not change or challenge existing roles and power structures will be accepted faster than those that dislocate the status quo
Legality	Ideas that are required by law or regulation will be accepted faster than those which are voluntary or ambiguous with existing laws and regulations
Cultural sensitivity	Innovations which are sympathetic to existing cultural norms, customs and habits will be accepted faster

(Adapted from: Spence 1994, Rogers 1995 and Baumol 2010)

to clarify objectives and set clear measurable performance indicators and to create a system for tracking, monitoring and reporting against them. In building the plan, resourcing and budgets must be carefully considered as should the establishment of a performance management system to motivate people to achieve it. An effective communication strategy should also be developed with different versions of the plan being produced for each key stakeholder group, setting up monthly strategy meetings with established reporting and reviews to monitor progress.

Confirmation

It is at the confirmation stage that innovators seek reinforcement and feedback from others about the effectiveness of the implementation process. There are only two outcomes from this stage: discontinuance or continuance. Discontinuance is the decision to reject an innovation which can occur for a variety of reasons. For example, the target population may decide to replace the innovation with a better idea that supersedes it or they may be rejected out of dissatisfaction with its performance. Alternatively, rejection might arise out of the misuse of an innovation, out of resource constraints to continue its use or from a failure to institutionalize it within on-going practices. Interestingly, research has found that later adopters are more likely to discontinue innovations than early adopters.

Measuring innovation success

The measurement of innovation output and success is a controversial topic. The standing joke among innovation managers is that ROI (Return on Investment) really stands for 'Restraint on Innovation' (Morris 2008). While on the one hand innovators need to be accountable for their actions like everyone else, on the other hand there is also evidence that the act of measurement can stifle risk taking and promote the type of short-term focus which is the antithesis of innovative thinking.

There are many challenges in measuring ROI from innovation. For example, how does one measure the return on investment from an idea which might produce results in ten years' time? The effect of an innovation might take many years to mature. Consider for example, the many decades of minute incremental developments in transistor technologies which eventually led to the development of the mobile phone, the digital camera or the personal computer. To not invest in an idea simply because its value cannot be measured today will inevitably mean that their value will never be realized. For example, if the scientist working for Canon in the 1970s, who accidentally dropped his soldering iron on his pen, had been forced to prove the ROI of his hunch that superheated ink could have potential uses in printing, the company may not have developed the first ink-jet printers and realized the benefits of an international multimillion dollar business. Opponents of measurement argue that many ground-breaking ideas like this have been abandoned by firms simply because they don't appear to produce immediate returns. Another example is the world's first re-usable space shuttle which was developed by a team of aeronautical engineers at the British Aircraft Corporation in the 1960s. The British government abandoned the project because of short-term budgetary concerns but three years later, the Americans became interested in the concept of a reusable spacecraft and eventually developed the idea into the Space Shuttle we know today. As Tom Smith, the leader of the British team of engineers, poignantly said, 'there is nothing worse than being right at the wrong time' (SMH 2012: 18). The history of ideas like this tells us that innovation is a venture into the unknown which might not produce immediate results. Indeed, truly disruptive innovations are often worse than existing incumbent products and services at the time of their launch. Consider for example the first digital cameras whose picture quality when introduced was far inferior to traditional film cameras.

Another challenge in measuring the ROI from innovation is the 'pipeline effect' that most innovations produce. While the first iteration of a particular innovation might not pay immediate results, a minor refinement which follows might do so 'in spades'. In other words, innovation is a 'process' not a one-off event and most good ideas build incrementally on each other over a period of time.

Finally, there is also the question of whether we measure outputs or inputs to the process. Simply measuring the 'output' of an innovation (in terms of profit or productivity) is likely to miss important 'input' measures which also reflect levels of innovation in a business. As we have seen in this book, failure is as important an indicator of innovation as success. Indeed, Daphne Kwon and Bill Hildebolt, CEO and President of ExpoTV.com (winner of the 2007 Stevie 'most innovative company' award for women in business), have developed specific business KPIs around failure.

Measuring innovation success in practice

Despite the dilemma of measuring ROI associated with innovation, most companies require their managers to measure and report on the extent and success of their innovation activities. Like most things in business, any initiatives which involve the investment of capital and time have to be measured and accounted for. However, in measuring innovation success many firms have never defined what innovation means to their business in terms of its desired contribution to business objectives. Furthermore, they more often than not fail to establish expected and measurable innovation performance indicators and to determine the best ways of measuring those outcomes. The problem, as Palangkaraya et al. (2011) point out, is that there is no one ideal all-encompassing measure of innovation activity. The most commonly used measures of innovation include: expenditure data (such as R&D); IP count-based data (such as designs, patent and trade mark applications); and qualitative innovation assessments (such as surveys of customer attitudes towards the firm). However, some measures (such as trade mark applications) are more suited to capturing product rather than process innovations while others (such as patents) are more suited to capturing radical rather than incremental innovations. Total (multi) factor productivity growth is also considered by some to be a good indicator of innovative activity. However, Hall (2011) warns that increases in productivity are by no means certain from innovation or indeed easy to measure if they are. For example, many process innovations will reduce costs (and increase profit) but not necessarily increase output. Furthermore, increases in productivity can at best only be treated as a proxy for innovation activity since it is inevitably related to other factors in the firm's environment. Finally, having focused so far on 'economic' measures of innovation success, OECD (2010) notes that many innovations address societal needs or may have spill-over effects for workers and communities. It is therefore important to develop new measures of innovation that address a range of impacts as well as economic ones. While accountants traditionally measure and report on only economic measures of success, advances in the field of corporate social responsibility have led to broader sets of measures such as triple bottom line accounting, social return on investment and shared value. There is a growing body of research which shows that innovative firms use broader measures of business performance than non-innovative firms and also have a wider variety of management controls, beyond traditional accounting systems, to measure and monitor them (Boedker et al. 2011). For example, a recent survey by McKinsey Consulting of 1016 company directors found that the information that they most wanted was, in order of priority (McKinsey 2005):

- Market health – customer satisfaction, competitor intelligence, suppliers and brands.
- Organizational health – employee retention, engagement, capabilities and skills, culture.
- Network health – community views, strategic relationships, government policies/regulation changes.
- Financial performance – cash, costs, margins, return on capital.
- Operational health – buildings, inventories, patents, production rates.

It is now estimated that over 40 per cent of investment decisions on financial markets are made on the basis of non-financial information and many leading firms now use much broader measures of success than purely financial information.

For example, some organizations base measurements around a balanced scorecard approach which recognizes that organizations should use a range of financial and non-financial indicators to measure success (Kaplan and Norton 1992). These measures typically include:

- Financial (new revenue streams, R&D spending etc.)
- Customer (measures of customer satisfaction and perceptions etc.).
- Internal business processes (time to market etc.)
- Learning and growth (patents, developing new markets etc.)

While widely used, the balanced scorecard approach has attracted criticism from a variety of sources. For example, some dislike the empirical nature of the framework while others argue that it treats goals in isolation and is simply a list of metrics which does not provide a bottom line score or a unified view with clear recommendations.

Businesses, within and outside the construction sector, have begun to adopt the concept of corporate social responsibility (CSR) which balances economic, social and environmental measures of success, guided by measurement frameworks such as the Global Reporting Initiative (GRI). However, as Loosemore and Phua (2011) found in their analysis of CSR in the construction sector, the rhetoric often doesn't match the reality and most firms are still very much driven by the bottom line. Furthermore, while advances have been made in a handful of leading construction firms in measuring environmental goals, the area of social responsibility remains a challenge.

More recently, in the most downloaded paper of 2012, Michael Porter and his colleagues at FSG have questioned the concept of CSR and developed the concept of 'shared value' by reviewing the effectiveness of social impact measurement approaches developed by companies like Nestlé, Intel and the InterContinental Hotel Group (Porter et al. 2011). However, despite enormous interest in the concept of shared value in business, tools to put the concept into practice remain in their infancy. According to Porter et al., effective measurement of shared value starts with a well-defined value strategy which identifies key social issues which align with business goals. Most companies do not have this in place. Having identified the key social issues to target and made the business case for doing so, shared value measurement then assesses and tracks progress against these goals, generating actionable data and insights to further refine shared-value strategies and unlock new value and so on.

Innovation data

Regardless of the method used to measure innovation, the quality of data collected is paramount to give a reliable indication of innovation success. Problematically, not all firms collect R&D data or report it and the impression gained can therefore be highly dependent on a firm's strategic motivations. Furthermore, since much of the data used by firms to measure innovation activity is unreliable and because measures of innovation activity are not yet widely established, the proclamations of many firms which claim they are innovators should be treated with a healthy degree of skepticism until properly investigated. For example, R&D data normally comes from accounting sources although is often inaccurate because of the lack of mandatory reporting standards and requirements in this area. IP counts are also problematic since they mostly apply to 'manufacturable' innovations and almost totally exclude organizational, service and market innovations which are common in the construction sector. Furthermore, IP counts such as patents are not always used by firms to protect their ideas. This is particularly the case in firms working in technical fields such as construction which are not covered well by patent laws and where firms are too small to have the resources to protect their patents. Furthermore, inventions can be protected by other methods such as secrecy, unregistered copyright and keeping ahead of the competition or simply by high barriers to entry into markets which ensures that innovations are hard to imitate. For these reasons, patents only tend to be a reliable indicator of innovation in manufacturing industries such as pharmaceuticals, machinery and precision instruments. So in most construction firms, because governments commonly use indicators like patents as a measure of innovation activity, innovations remain invisible, uncategorized and indistinguishable from other company expenses. While customer and manager surveys can address these problems to some extent and produce much broader measures of innovation activity, the quality of many innovation surveys is questionable and results can be affected by subjectivity, non-response bias, sample selection and failure to consider non-responses and innovation quality. Measures of numbers of new product or service launches can also be useful, but again they are generally unable to distinguish between new and imitated products. Finally, while publication citations might indicate levels of innovation activity, they can often be of questionable quality (which publications really matter?) and in any event, many firms have a direct interest in maintaining secrecy about their new ideas and avoiding citations in journals.

Dodgson and Hinze (2000) argue that multiple indicators should be used to make any judgements about levels of innovation and to this end, Morris (2008) proposes an alternative way to measure innovation success which sees the innovation process as a 'funnel'. This starts broadly with strategic and creative thinking about lots of ideas which gradually focuses over time on idea selection and development, implementation and eventually sales. The advantage of this approach is its recognition that different measures of innovation are appropriate at different stages of the innovation process (see an adaptation of Morris's metrics in Table 2.4).

International measures of innovation success

A growing recognition that innovation is a significant determinant of economic well-being and prosperity has led to the development and publication of a number of innovation indexes which compare countries on the basis of their innovativeness. One of the most widely respected was developed by INSEAD and the World Intellectual Property Organization (a specialized agency of the United Nations) and is called The Global Innovation Index (GII). Published annually as an international comparator of national innovation activity, the top ten countries in 2012 were:

1. Switzerland
2. Sweden
3. Singapore
4. Finland

5. United Kingdom
6. Netherlands
7. Denmark
8. Hong Kong (China)
9. Ireland
10. United States of America.

Table 2.4 Innovation metrics at different stages of the innovation process

	Stage	*Qualitative measures*	*Quantitative measures*
INPUT	Strategic thinking (linking innovation to corporate objectives)	Do our innovation goals align with corporate goals? In what areas do we want innovation to happen and for what reason? Do we know how to measure innovation success? Are our metrics evoking the sorts of behaviours we want from people in our business? Are we capable of innovating in response to changes in strategies initiated by market changes?	Time senior managers invest in innovation? Resources invested in innovation activities? Time from strategic development to sales? Target ROI from innovation?
	Innovation management (managing diversified innovation projects)	Do we have an innovation plan (with clear objectives)? Do we have the right balance of incremental and breakthrough projects? How are we tracking the development of different projects? How does our portfolio cover our market segments? How does our portfolio compare with our competitors'? How are resources spread across the portfolio of projects and is it appropriate? Do we need different metrics in each area of portfolio to measure ROI?	Time senior managers invest in innovation in each area? Resources invested in innovation activities in each area? Time from strategic development to sales in each area? Target ROI from innovation in each area?
PROCESS	Research (generating new knowledge by exploring problems and digging for unknown and unmet needs)	How well do we understand the future (technologies, customers and markets)? Is our research exposing future innovation opportunities? Do we have a research plan/ roadmap (with clear objectives)? Who is doing our research (quality/bias/breadth)?	Number of research projects (sample sizes/scope etc.)? Breadth of participation throughout the organization? External grant successes? Time invested in research? Money invested in research? Balance between blue-sky and applied research? Alignment of research project

Table 2.4 continued

	Stage	*Qualitative measures*	*Quantitative measures*
		How is our research being managed?	outcomes with innovation objectives? Numbers of new insights produced? Dissemination of new knowledge (internally/externally)? External recognition?
	Ideation (generating new ideas by combining knowledge from different domains)	Do we have a culture of ideation (where people can take risks and instinctively share knowledge and ideas)? Do we have silos? Are people territorial with information? Do we have systems in place to 'gather' ideas? How easy is it for ideas to emerge from different parts of the business?	Number of new ideas generated? Number and range of ideation forums? Breadth of involvement in ideation forums (participation)? Percentage of ideas from outside? Percentage of ideas from inside?
	Insight (selecting the good ideas that match needs and market opportunities)	Are we able to spot good ideas that have market potential? How broadly are these solid ideas being developed?	Actionable insights as a proportion of ideas generated? Number of new insights taken forward to development? Number of ideas implemented?
	Targeting (aim the good ideas at specific market needs)	What is our ability to develop our insights to specifically meet strategic objectives and market needs? Is our portfolio of insights balanced correctly?	Number of ideas funded for full development? Number of ideas killed? Percentage of ideas in core and non-core areas?
	Development (making the ideas work - transforming them into new successful products and services)	Do we have the project management skills and resources to develop-up our ideas through design, prototyping, testing, refinement, business modeling/planning? Are the right people involved in this process? Do we have enough failures to ensure we are pushing the envelope?	Innovations ready for market? Number of successes v failures? Speed of development? Numbers of patents applied for?
O U T P	Market development and selling (marketing,	How well do we understand our customers (existing and new)? Do we have the skills and	Number of new products or services launched? Time to market from development to sales?

	Stage	*Qualitative measures*	*Quantitative measures*
U T	selling, delivering and servicing the new ideas)	knowledge to convert ideas into buyable products and services? How well are we balancing our attempts to reach new versus existing customers? Do our customers value our products and services? What is our brand and reputation among our existing and new customer bases?	Average age products and services? Average life span of products and services? Return on marketing investment? Growth rate of customer base (numbers and breadth)? Profit growth? Gross sales revenue per innovation? Gross sales margin per innovation? Sales growth per innovation? Benefits and spill-over effects of innovations (internal, customers, society etc.)? Customer satisfaction?

(Adapted from Morris 2008)

The Global Innovation Index (GII) is based on two sub-indices: the Innovation Input sub-index and; the innovation output sub-index.

The innovation input sub-index factors include:

- Institutions (political, regulatory, business environment)
- Human capital and research (education and R&D)
- Infrastructure (ICT, general infrastructure and ecological sustainability)
- Market sophistication (credit, investment, trade and competition)
- Business sophistication (knowledge works, innovation linkages, knowledge absorption).

The innovation output sub-index factors include:

- Knowledge and technology outputs (knowledge creation, knowledge impact, knowledge diffusion)
- Creative outputs (creative intangibles creative goods and services, online creativity).

One of the criticisms of this index is that it doesn't include references to national culture.

The other major national innovation index is produced by the Economist Intelligence Unit and ranks 82 countries based on their innovativeness. The top ten countries in 2012 were:

1. Japan
2. Switzerland
3. Finland
4. US
5. Sweden
6. German
7. Taiwan
8. Netherlands
9. Israel
10. Denmark

The Economist Intelligence Unit measures innovation output by the number of patents granted to people from different countries by the patent offices of the US, the European Union and Japan. The index also looks at factors that help or hinder the ability to innovate, such as the amount of research and development (R&D) and the technical skills of the workforce. There are also two composite indices, constructed on the basis of innovation inputs, or innovation enablers. The factors included under the innovation inputs index are:

- R&D as a percentage of GDP
- quality of the local research infrastructure
- education of the workforce
- technical skills of the workforce
- quality of IT and communications infrastructure
- broadband penetration.

The factors included under the innovation inputs enablers are:

- political stability
- macroeconomic stability
- institutional framework
- regulatory environment
- tax regime
- flexibility of labour market
- openness of national economy to foreign investment
- ease of hiring foreign nationals
- openness of national culture to foreign influence
- popular attitudes towards scientific advancements
- access to investment finance
- protection of intellectual property.

In response to these indexes and to increasing global competition, individual countries are developing their own national innovation strategies. For example, in Australia, the Culter Review 'Venturousaustralia' (2008) called for a significant recasting of Australia's innovation policy by giving priority to building stronger partnerships between firms, universities and publically funded research agencies and by building the capacity of firms to absorb and incorporate new knowledge. Informed by this review, the Australian Federal Government outlined its innovation agenda for Australia over the next decade in *Powering Ideas: An Innovation Agenda for the 21st Century* (COA 2009). *Powering Ideas* outlined the Australian Government's plans, priorities and targets for its national innovation system with a focus on: skills and research capacity, business innovation, links and collaboration and public sector innovation. Its 72 recommendations ranged across the components of an effective system, focusing on innovation in business, strengthening people and skills, excellence in national research, information and market design, and taxation. *Powering Ideas* outlines a vision for a national innovation system in 2020 which includes: increasing the number of Australian research groups performing at world-class levels; boosting international research collaboration by Australian universities; significantly increasing the number of students completing higher degrees by research over the next decade; doubling the level of collaboration between Australian businesses, universities and publicly funded research agencies; a 25 per cent increase in the proportion of businesses engaging in innovation; and continued improvement in the number of businesses investing in R&D.

In 2010 the OECD also launched its innovation strategy to help governments design and coordinate their approaches to finding national and global solutions to global challenges such as climate change, poverty and food security. The strategy is based on the belief that such challenges require a coordinated effort. Other regional initiatives include the EU KLEMS project on growth and productivity in the European Union which compared industry productivity across EU member states, Japan and US from 1970 onwards. Globalization is also leading to regional accords like the Lisbon Strategy which was designed to address low productivity and economic stagnation in the EU through various non-binding policy initiatives. The Lisbon Strategy presents a 10-year plan for reviving the EU economy through the promotion of 'smart, sustainable, inclusive growth.' The aim is to increase investment in R&D by 3 per cent of GDP, raise employment rates, reduce greenhouse gas emissions, increase renewable energy production, increase education and reduce poverty.

Conclusion

The aim of this chapter was to better understand and critically appraise the concept of innovation. It is evident that there are many different types of innovation and that innovation is a complex and multidimensional process which occurs in different ways in different industries. To pigeon-hole all innovation across all industries into traditional linear scientific models is clearly inappropriate and new methods of measurement are required to reflect this diversity. Traditional measures of innovation underplay the real levels of innovation that occur in the construction sector. In industries like construction, innovation arises from new ways of thinking, interacting and communicating and does not necessarily require the large, high-technology, resource-intensive R&D type initiatives which traditional notions of innovation measure. Greater levels of innovation will require the shifting of functional and intellectual boundaries between the different professions and knowledge domains within the construction sector. Firms will also need to reassess and more clearly define their role in the new value streams that emerge in the future.

3 Innovation in construction

If the only tool you have is a hammer, you tend to see every problem as a nail.
Abraham Maslow

Introduction

In many parts of the world, serious attempts to reform the construction industry go back to the 1960s but gathered momentum during the 1980s and 1990s in response to accumulating evidence that the construction industry's performance was unreliable and that its production processes were unsafe and failing to keep pace with technological advances that were revolutionizing society. Cost overruns, program delays, conflict and confrontation were the norm for customers of the construction industry and in response to the industry's apparent inability to reform itself, major clients and governments around the world took it upon themselves to instigate reform from the outside. During this period of unprecedented reform, many reviews of the industry's performance were commissioned, alternative forms of procurement and contract were produced and industry-wide reform bodies were established. Among the most active and influential were Construction Excellence in the UK, Construction Industry Development Agency (CIDA) in Australia, the Building and Construction Authority (BCA) in Singapore, the Construction Industry Institute (CII) in the USA and the Construction Industry Council (CIC) in Hong Kong.

Some authors have even questioned the faulty reasoning on which the need for reform was based in the first place. For example, in the UK, Edkins and Winch (1999) concluded that there was no evidence that UK buildings were relatively expensive during the 1980s compared to other countries and that relative prices were linked strongly to economic conditions for example, when Sir Michael Latham's influential report *Constructing the Team* (1994) was commissioned in the UK, construction costs were relatively high due to a building boom and a strong pound. However, when the Latham Report was published, costs were relatively low due to a recession and weak pound. Furthermore, there was a tendency towards higher quality specification in the UK, compared to other countries which distorted international comparisons. And it was also found that labour productivity was relatively low in the UK because there was a design culture which favoured bespoke solutions.

In many cases, the proposed solution to the construction industry's problems has been the unthoughtful transplantation of manufacturing techniques into the sector. But Winch (2008: 23) has questioned the continued 'holding-up' of manufacturing as best practice pointing out that the automobile sector also 'has its incompetent back-street garages and dodgy car salesmen'. Winch argues that it is important to recognize that the construction sector at its best displays highly sophisticated skills in addressing highly complex and difficult production problems in a way that draws considerable respect from other industry sectors. Furthermore, as the many prominent contributors to Dainty and Loosemore's (2012) critical review of human resource management practices in the industry shows, the widespread adoption of popularized management techniques such as partnering, total quality management, lean production, partnering, benchmarking and re-engineering have had mixed, if any, success. Indeed, many have been counterproductive.

Is the construction sector less innovative than other sectors?

Given decades of damning reports on the relative performance of the construction industry compared to other sectors such as manufacturing, the automatic temptation is to answer this question with a resounding NO. For example, the Australian Innovation System Report (DIISR 2010) claimed that the construction sector has historically had low levels of innovation with only 30.8 per cent of businesses innovating (the lowest of any sector) and this has been widely quoted to support this negative perception (see for example Australia's recent National Building Information Modeling Initiative – SMART 2012). However, it would seem that the construction industry is not as 'backward' as authors like Woudhuysen and Abbey (2004) would have us believe and that any debate over the relative performance of the sector is more complex than it might first seem. Like all industries, there is undoubted potential for greater efficiency in the construction sector, but as Gambatese and Hallowell (2011) point out, too many studies lament the industry for being resistant to innovation and change, providing fertile ground for people to explore the reasons why this may be so, leading to a self-fulfilling philosophy that this must be the case. A more informed and correct answer to the above question must rest on an intimate understanding of the differences between manufacturing and project-based industries like construction and of the industry's culture, history, structure, processes and ways of working. For example, in contrast to manufacturing which is typically technology intensive, construction is a service-based industry which is inherently labour intensive. While construction is becoming more technology-driven, innovation is fundamentally more 'behavioural' than R&D based (Sundbo 1996). Furthermore, as Barrett et al. (2008) point out, the vast majority of innovation theory has been derived from R&D studies of manufacturing firms. And most of the research on innovation has been focused on large enterprises and neglects to sufficiently consider the special innovation challenges faced by the many small-to-medium-sized enterprises (SMEs) that dominate the construction industry (Sexton and Barrett 2003, 2005). Sexton et al. (2008) found that SMEs are motivated to innovate by different factors to large firms (survival rather than growth and by the need to solve immediate project-related problems). SMEs are also more open and responsive to their market environment than large firms and rely more on personal client relationships and on the tacit knowledge and experience of their employees. Typically, they are also less technology-intensive, lack the slack resources to innovate and are more highly geared than large firms.

Barrett et al. (2008) list numerous other important differences between construction and manufacturing which might cause us to carefully question any simplistic comparisons of innovation. For example, when one looks at the construction sector as a system of contributors over the whole life-cycle of a building, it is clear that a much larger constituency of stakeholders need to play a role in the innovation process. Within this social context there are also politics to be handled and vested interests to be considered. And in contrast to manufacturing, innovations in the construction industry are rarely large scale and radical but small-scale incremental improvements in services or products. Innovation in construction also tends to be more ad-hoc than in manufacturing firms, based on ideas from employees and managers developed incrementally 'along the way' in response to challenges during the service delivery process. As Gronroos (2000) shows, even when a service-based organization develops a systematic and planned innovation process, innovations themselves must be developed by the people working in that organization, often working closely in collaboration with customers as co-producers of knowledge. People with ideas then have to fight for the recognition of their ideas and the business culture, climate and structure must enable these new ideas to rise to the surface otherwise these ideas will die and become wasted and lost.

Our understanding of these differences and what they mean for innovation is only starting to emerge. But we do know that to rely on traditional measures of innovation to compare the construction industry's performance with other industries is potentially deceiving. For the reasons above, and many others we explore in this book, innovation in the construction industry does not happen in the way that traditional R&D-based measures pick up. Few firms in the construction sector generate radical new ideas in the laboratory but incrementally at a project level in solving practical day-to-day problems. Many firms in the construction industry excel at innovating at this operational project level. And it can be argued that while these companies might not invest in speculative, experimental and laboratory-based experimentation, they are actively innovating nonetheless.

While the portrait of innovation in the construction industry may not be as bleak as its many critics would suggest, this is not to say that it cannot be much improved. There are clearly many institutional, cultural and organizational barriers to innovation which need to addressed to help managers in the industry build the innovative and productive capacities of their firms. And it is to these that the rest of this chapter is dedicated. We start by exploring the controversial role of clients in the innovation process.

The role of clients in the innovation process

Cutler (2008) makes the point that while there are many similarities and differences between innovation in manufacturing and service-based industries, what crucially distinguishes service-based innovation is the central role of customers in the innovation process. Certainly the important role of the customer has been widely recognized in the construction sector for many years. As far back as Sir Kenneth Wood's influential report into the role of *The Public Client and the Construction Industry* (Wood 1975), it has been appreciated that commissioning a building construction is not the same as purchasing a typical consumer product. It is an infinitely more complex and high-value purchase and innovation has a better chance of success if customers and clients are intimately involved in the development of new ideas from the very start of the process. This view is also supported by Gambatese and Hallowell (2011) whose research into the factors that influence innovation in the construction sector pointed to the importance of client support in valuing and driving innovation. Finally, Egbu (2008) and Barrett (2008) argue that clients can come together and through their combined purchasing power help the industry create a consensus around a meaningful shared vision and a way to implement it. Clients also have a major role to play in the way that the construction market operates and they can aspire to be better clients by demanding innovation, by avoiding cut-throat competition which undercuts margins and by creating a trusting and stable environment for innovation to occur through equitable risk distribution and by taking a longer-term 'value-driven' view of their building investment over its entire life-cycle rather than seeing it as a short-term 'construction cost'.

While many point to the crucial role of clients in promoting more innovation in the construction sector, Brandon and Shu-Ling (2008) argue that relying on clients as drivers of innovation is a 'cop out' by the industry. Relying on clients to drive innovation, they argue, discourages firms from investing in their own new ideas and from recognizing and acting out their shared responsibility for the advancement of the industry. This position is supported by Hobson and Treadaway (2008: 26) who argue that while clients should be able to clearly communicate their needs to the building team and make timely decisions etc. – 'there is no reason why the client should be expected to know anything about construction . . . So why are we expected to help the construction industry with its internal workings?'. Indeed, Bresnen (2008) argues that there is a fundamental problem in assuming that clients have a coherent and uncontested agenda when it comes to innovation. Often the main goals of innovation from a client's perspective is to simply emulate their competitors' success notwithstanding whether the proposed innovation is in its own interest or not. Also, as Tzortzopoulos et al. (2008) argue, attempts by clients to introduce new management initiatives like partnering etc. may be motivated by selfish rather than positive intentions – such as being able to keep prices down during prosperous times when competitive tendering would otherwise force prices up. For example, it has been widely reported that many major clients in the UK construction industry who championed partnering during the 1980s as a new collaborative way to do business hypocritically jettisoned this approach when the market changed and it suited them to revert back to lowest price competitive tendering practices.

Regardless of one's views about the responsibility of clients to drive more innovation in the construction industry, there is plenty of evidence to indicate that their role is crucial and that, in most instances, they may be a barrier to innovation rather than a help. For example, Loosemore and Phua's (2011) research into the industry's CSR record found that many clients are simply unprepared to pay for more socially responsible investments, seeing little value for their core business and being ill-equipped to measure its value. Loosemore and Phua's research showed that for many aspiring firms in the construction sector, clients represent a formidable barrier to innovation and often have to be 'worked around' by quietly introducing innovations into their projects without their knowledge. These findings support those of Sexton et al. (2008) who found that while some prominent clients take their leadership responsibilities seriously, the majority of construction clients play a passive role in the innovation process.

According to Ivory (2005), the main reasons why clients are unwilling and unable to effectively encourage innovation are:

1. Buildings are often seen as a tradable short-term asset rather than a long-term strategic investment. Many clients do not understand the relationship between their buildings and their core business performance.
2. Competition with other internal core business priorities.
3. High short-term capital costs involved in innovating.
4. Internal politics and difficulties in achieving consensus between project stakeholders.
5. Increased risk exposure and workloads in managing innovation.
6. Indemonstrable benefits from investments in innovation.
7. Inadequate competence to confidently manage the process and judge net benefits.

Of course, one cannot generalize about all clients. As Manley (2006) points out, the construction sector serves a wide variety of clients with an equally wide variety of needs and not all clients are equally mature in terms of their approach to the innovation process. Manley's research concluded that large repeat clients are among the most likely to drive innovation in the construction sector, followed by governments and then one-off clients. However, it is clear that the maturity of clients' attitudes towards innovation will also vary from sector to sector with some clients imposing their own high expectations of innovation onto their business partners. For example, when Australia's oldest family-owned builder, Kell and Rigby Ltd, accepted the contract to build Apple Stores throughout Australia they also accepted the challenge of working with one of the world's most innovative companies and their procurement approaches had to reflect this culture.

Lim and Ofori's (2007) research may help explain this phenomenon because it recognizes that not all construction clients will pay for innovation and that not all innovation costs can be passed to clients. They argue that there are three types of innovation in construction:

1. Type-1 innovations – are those which clients are willing to pay for.
2. Type-2 innovations – are those that reduce construction costs or improve productivity.
3. Type-3 innovations – are those that produce a sustained competitive advantage through differentiation.

Lim and Ofori argue that to thrive off Type-1 innovations a firm must work in close long-term partnerships with their clients in order to continuously and deeply understand the needs, business imperatives and the role that buildings play in their client's core business success. This will allow them to identify the right demand signals to respond to. Those firms which are unable to engage in close partnerships with clients are limited to pursuing Type-2 innovations to produce price advantages in the market. All firms should pursue Type-3 innovations which create opportunities for extraordinary profits from completely new ideas which differentiate them in the market.

Sexton et al. (2008) also challenge the illusion of the all-powerful client's role in driving innovation. They argue that a client's role in the innovation process varies from 'passive' to 'balanced' to 'dominant' depending on the attributes of the client involved. Like Manley et al. (2006) they found that dominant clients actively engage with driving innovation in the industry and tend to be those which repeatedly use the industry to procure assets critical to its core business. These types of clients therefore have a central interest in directing the way those assets are built. In contrast, the passive client is normally a small, inexperienced, speculative or one-off user of the industry that tends to consume off-the-shelf products. These clients prefer to use the same standardized products and services over and over again and operate at arm's length from the industry. Finally, between these two extremes is the type of client which works in partnership with the industry to develop innovative solutions to their needs. These clients tend to be large, intelligent and informed businesses which have a clear vision for what they want. They need not be repeat users of the industry but they know about managing projects successfully. Sexton et al. argue that this type of client has been the underlying assumption for most models of innovation in the construction management literature.

In extending our understanding of clients and their role in the innovation process, Tzortzopoulos et al.'s (2008) classification framework is useful. Their client categories are defined and listed in Table 3.1 with some speculation as to their role in the innovation process (although there has as yet been no empirical research to test these propositions).

While typologies are useful, Tzortzopoulos et al. (2008) acknowledge that they do not address individual client needs and motivations to innovate. In reality firms need to develop a relationship with each and every client to develop an understanding of their needs. Only then will they be able to identify the innovative ideas that might be able to add value in the marketplace.

Table 3.1 Types of clients and their role in the innovation process

Client type	*Description*	*Role in innovation process*
Identifiable/ unitary	Easily identifiable and tend to have specific and easily identifiable design problems. They have a clear view of their needs which are able to be discussed directly with them at a single interface (e.g. family needing a new house).	Are likely to be passive and risk averse and to invest in limited innovations offered off the shelf. There is a single point of responsibility and authority for decision-making which makes decisions about innovations easier.
Virtual/ pluralistic	Multifaceted and complex with a range of needs coming from an array of internal stakeholders whose interests may not be clear or easily reconcilable. Innovators have to deal with multiple interfaces.	May be open to innovation but internal conflicts and inconsistencies in understanding and agreeing on needs may make agreement and implementation of innovations difficult.
Paying clients	An identifiable customer signing off payments and documents during the building procurement process.	A single point of responsibility and decision-making makes it easier to invest in innovation but only when benefits are evident. Separation from users or beneficiaries of innovation may limit ability to recognize these.
End users	Individuals or organizations which use the built facility who may not be identifiable during the building procurement process.	Unlikely to be in a position to invest in innovation since do not control up-front resources or get involved in up-front decisions. However, they are in a good position to comment on any potential benefits and work with the industry to innovate as potential beneficiaries.
Sophisticated	Experienced repeat users of the industry which understand the internal workings of the industry and procurement process. If they	Are likely to be active supporters of innovation and are most likely to work in collaboration with the industry to develop new ideas.

Table 3.1 continued

Client type	*Description*	*Role in innovation process*
	are large they may have internal teams to manage the process.	They can drive innovation through large project programs and can make fast decisions in implementing them. Are also likely to have the capacity, resources, expertise to drive innovation and the ability to benefit from it over multiple projects.
Naive	Partially or uninformed clients with little experience of the industry or of the building process. Are likely to need advice and guidance on how to commission buildings.	Unlikely to understand the benefits of innovation or the reason for it. But may be open to investing in innovation if benefits can be explained clearly.
Primary	Income derived from constructing buildings (developers).	Are likely to be active supporters on innovation and most likely to work in collaboration with the industry to innovate but only on the basis of a strong business case and obvious benefits for efficiency.
Secondary	They build as a means to perform another core business activity.	If benefits for the secondary activity can be demonstrated then are likely to be open to considering and supporting innovation.
Continuing	Have large on-going construction portfolios and repeat users of the industry.	Are likely to be active supporters on innovation and most likely to work in collaboration with the industry to innovate since benefits can be realized and multiplied over many projects.
One-off	Involved in one-off projects and infrequent users of the industry.	Unlikely to understand the benefits of innovation or the reason for it. There is an expectation it is provided as part of the service. Prefer off the shelf innovations rather than to collaborate to develop them on a one-off, individual project basis.
Private sector	Requirements are normally to meet economic needs (e.g. industrial or commercial)	Likely to invest in innovation if business case is proven.
Public sector	Requirements are designed to meet social needs and are seldom clear cut. High accountability and	Through their control of large work portfolios, have the leverage and capacity to drive innovation

Client type	*Description*	*Role in innovation process*
	process driven (can be national or local).	on large projects and multiple projects and lead the industry as best practice clients but are likely to be constrained by risk averse, cultures of governance, slow decision-making and accountability.
Mixed client	NGO, not-for-profit etc. – have both social and economic imperatives.	Likely to support innovation if satisfy social and economic imperatives although may have limited means and resources to do so.
Stable environment clients	Clients operate in a stable business environment and their needs are also stable during design and production.	Are unlikely to invest in innovation since they operate in stable business environments where innovation is not a premium. Decisions to invest in innovation are likely to be few and far between but backed up by certain support and commitment.
Dynamic environment clients	Clients operate in a dynamic business environment and their needs are likely to change as design and production evolves.	Will place innovation at a premium in helping to secure competitive advantage and respond to ever changing business needs and environment. But nature and benefits likely to change over time, making innovations a moving target and speed of implementation critical.

(Adapted from: Tzortzopoulos et al. 2008)

The role of competition in driving innovation

The issue of whether competition is good or bad for innovation in the construction sector is a contentious one. On one hand competition can stimulate innovation by encouraging companies to 'out-do' each other. On the other hand, when the primary objective of competition is to secure the lowest price at any cost, it can be highly counter-productive to innovation by undercutting the margins and slack resources which are crucial to investments in new ideas. Both sides of the argument were put by the Cole Australian Royal Commission into the Building and Construction Industry (RCBCI 2002a). The Final Royal Commission report noted that there was virtually no international competition in the Australian construction industry, few alternative products for clients to choose from and significant barriers for new firms wishing to enter the industry. Therefore, it was argued, new players which are likely to be more innovative are excluded from the industry and there is little pressure on existing firms to break rank and run the risk of trying something different. While there is undoubtedly some truth in these arguments, the industry is highly cyclical and when times are bad, fierce domestic competition for work becomes an important factor in driving innovation, particularly during the early stages of a project when firms are competing for new work. Furthermore, while competition is good for innovation, there is also a point when excessive competition becomes destructive by forcing down margins to untenable levels. This potential problem was also recognized by the Australian Royal Commission which noted that the tight profit margins in the industry have led to head contractors seeking avenues to reduce risks wherever possible by passing them off to subcontractors who often are unaware of or incapable of managing those risks effectively (RCBCI 2002b). Too often, the ironic result is that clients, in an attempt to save money in the short term, suffer in the long term as their project descends into conflict and costly dispute.

In contrast to the normal experience of construction which is described above, what is needed for innovation is a trusting, open, risk-sharing environment which fosters cooperation.

Procurement systems and innovation

Procurement systems influence innovation because they set the parameters for competition, resourcing, risk management, collaboration and knowledge sharing on projects. As we discussed above, many construction companies, particularly in the major projects sector, have attempted to overcome the problems of excessive competition by nurturing close and long-term relationships with their clients (Nam and Tatum 1992). The intention of these partnership arrangements is to eliminate competition and to facilitate trusting relationships. The theory is that this then leads to a more consistent work-stream underpinned by more realistic resource levels, project development periods and more open, flexible, responsive and equitable contractual practices. As Nam and Tatum (1997) found, a high level of owner participation in project development and management fosters innovation particularly where there is a long-term relationship with designers and contractors. This creates a stable and predictable environment of collective responsibility, where there is a reduced risk of failure associated with innovation. Furthermore, parties have the time to cultivate relationships and understanding and thereby build confidence to initiate new ideas with the certainty of being able to benefit from them.

Unfortunately, the majority of construction clients are unsophisticated, procure buildings infrequently and find it difficult to see any direct benefit from reducing their emphasis upon price competition. This is the economic dilemma of reform in the construction industry. In reality, the potential for investment in innovation is restricted to the relatively few large companies who are lucky enough to deal with sophisticated and relatively wealthy clients which procure buildings on a frequent basis. The vast majority of the industry are left to work with clients who procure buildings very rarely, who want the lowest possible price for their investment and who do not see them as a key long-term asset in the success of their core business.

An alternative solution to forming partnerships with clients is to form alliances with competitors to develop new ideas from which both can benefit (Johnson and Tatum 1993). This strategy, known as 'co-opetition', has the advantage that the risks of innovation are also shared. However, Loosemore et al. (2002) found a strong view in the Australian Building and Construction Industry that competition means working 'against', rather than 'with' one's competitors. The idea of inter-firm cooperation to spread the risk of innovations is only likely to occur on mega-projects which are beyond the capacity of individual companies (in other words, out of necessity rather than innovation). The importance of collaboration to the innovation process in construction has been highlighted by numerous authors including Leiringer (2006) and Walker and Rowlinson (2008). They have pointed to the potential of collaborative procurement models (such as Alliances and PPPs) to stimulate innovation by building trust and teamwork between project members. However, despite the rhetoric of 'partnership', other researchers have shown that in many projects, they are simply a mechanism used by clients to transfer risk to the private sector (Akintoye et al. 2003; Dixon et al. 2005; Akintoye and Beck 2009). In other words, they are alliances or partnerships in name only. Furthermore, the alliance approach has had mixed success and is certainly not a solution to all types of building projects. Indeed, imposing an open-book approach on an industry where the majority have been steeped in a culture of confrontational working can be counter-productive rather than beneficial (Reed and Loosemore 2012). In the case of PPPs, it has also been argued that the high risk associated with innovation and the lack of risk sharing forces the PPP consortium down safe tried-and-tested routes rather than encouraging them to experiment with new innovative approaches. PPPs have been particularly contentious in areas like health where key industry stakeholders have argued that it generates the wrong type of innovation because the true natural drivers of innovation are the self-interested concerns of speculators whose objective is to minimize costs and time rather than deliver a better service to the customer (Smith 2006). While there are undoubted advantages of PPPs from an innovation perspective, such as encouraging a more integrated and life-cycle approach to design, construction and operation, the rhetoric and reality of 'partnership' working can be quite different in theory and reality.

The project-based nature of construction

Another barrier to inter-firm cooperation is the project-based nature of construction activity (Salter and Gann 2005). As Loosemore and Phua (2011) point out, in construction 'the project is king'. The project-based nature of construction inhibits the spread of innovations from one project to the next, ensuring that any diffusion remains limited to isolated pockets (Winch 1998). Long-term investments in innovation are unlikely to be amortized across projects and to furnish returns in time for those who invest in them. Loosemore et al. (2002) argue that this general lack of cross-fertilization of ideas between projects occurs because they are commonly treated as independent cost-centres and the uniqueness of each project and the lack of standardization provide little scope for transferring innovations. Furthermore, the increasingly tight time-scales and resource constraints being imposed upon project teams provide little opportunity for them to innovate. Consequently, any innovation that takes place does so in a reactive rather than proactive way and is confined to the duration of one project only.

While the project-based nature of construction activity appears to present a formidable barrier to innovation, Barrett and Lee's (2004) research into the perceived barriers to innovation in the construction sector ranked it last out of fourteen factors. Their findings ranked barriers to innovation in the construction sector in the following order:

1. Industry clients protecting their own interests
2. Industry bodies protecting their own interests
3. Lack of awareness of improvement initiatives
4. Lack of education and training
5. Lack of R&D
6. Lack of innovative procedures
7. Lack of technical innovation implementation
8. Lack of clear benefits
9. Extensive inter-organizational change required
10. Professional indemnity insurance cover
11. Lack of government leadership
12. Inappropriate legislation
13. Belief that industry is doing well enough
14. The temporary nature of projects.

Indeed, Dodgson et al. (2005) argue that project-based organizations can be highly conducive to innovation because they exhibit similar properties to Burns and Stalker's (1961) organic form of organization, offering the flexibility for people to experiment with new ideas and avoiding the rigidity of more routine type work and environments. The problem, according to Dodgson et al. (2005) is not projects themselves but the challenge in realizing these benefits by establishing a way for independent project-based organizations to exist alongside more traditional divisional or functional structures in a way which allows the benefits of innovation to be learnt and transferred from project to project.

Fragmented supply and demand chains

Building production takes place through a fragmented supply and demand chain which physically and chronologically separates customers from designers,

designers from builders and builders from operators. On most projects, these chains are also 'legally fragmented' by the way risks and rewards are distributed rather than being shared between project participants. According to Winch (2008) each firm in the construction supply chain competes to extract the greatest share of the value-stream flowing from the client. In theory this should be shared in proportion to their contribution to that value-stream process. However, Loosemore (1999) found that power differences between the parties ensures that in most projects value may not be extracted proportionately on this basis and this often results in confrontational relationships and a lack of collective responsibility for project success. Tatum (1987), Lamborde and Sanvido (1994) and Gann (2000) argue that this is one of the construction industry's major barriers to innovation – the separation of potential innovators from the potential benefits of their innovations. The result is that innovators further down the supply chain are, more often than not, separated from the final product into which their new idea is incorporated. For example, a domestic subcontractor's lack of involvement in design means that any technological innovations must be offered through a principal contractor who is likely to filter the potential benefits to the innovating subcontractor.

Another problem with construction is the lack of integration of product/material manufacturers into supply chains (Slaughter 1983). However, Gann (2000) argues that as construction becomes more manufacturing-based (through pre-fabrication and off-site manufacturing), traditional boundaries between manufacturing and construction project management services will become obsolete. In this emerging context the challenge for firms in the construction sector is to develop new integrated systems solutions which provide customers with a single point purchase and on-going support for buildings and structures. To some extent, procurement systems such as alliances, design and build, BOOT and PPPs address this need. These approaches incentivize integration and collective responsibility along the supply chain for design, project and building performance and move all project participants closer to the end product. So too do new technologies like business information modeling which, in theory at least, have the potential to facilitate greater integration of design, production and operational processes. However, the effectiveness of these new procurement systems and technologies is currently limited by the historic divided and confrontational culture of the construction industry (Walker and Lloyd-Walker 2011). As Reed and Loosemore (2012) point out, ultimately, alliance contracts like any other project are staffed by people who have been educated and instilled in the confrontational and fragmented traditions of the construction industry. This means that cultural assimilation and induction programmes will often be needed to change deeply imbedded ways of thinking (schemas) before people start working on such projects. Furthermore, given people's natural tendency to revert quickly back to instinctive learnt behaviours in response to the uncomfortable culture shock that will inevitably be experienced, programmes will need to be in place to monitor those behaviours and keep them changed.

Institutionalized impediments to innovation in the construction industry

The size and scope of the construction industry in most countries ensures that it tends to be afflicted, more than most sectors, by a complex, cluttered and fragmented regulatory environment. In Australia for example, there are over seventy overlapping bodies within the industry with different and often conflicting interests and agendas. As the Cole Royal Commission into the Building and Construction Industry noted, these are unlikely to come together and agree on an agenda to bring about meaningful change within the industry (RCBCI 2002a).

The role of government in driving innovation

Central to the construction sector's institutional network are governments which, as Fairclough (2002) pointed out, play a central role in the innovation process as a regulator, sponsor, client and policy-maker. However, with worldwide trends towards the privatization of public assets and an associated loss of skills and knowledge to lead the industry to innovate, the public sector's role in driving innovation has retreated to being largely regulatory and policy directed. In many countries, government policy-makers work at a considerable distance from those in the industry and public sector employees increasingly depend on the advice of outside consultants to provide them with new innovations and ideas. Therefore, in most countries, the knowledge and capability to drive innovation largely resides in the private not the public sector. Nevertheless, governments can still use a range of direct and indirect instruments to stimulate innovation where there has been a market failure to innovate. These include:

- R&D tax concessions
- Regulations
- Guidelines and policies
- Mandating the use of certain innovations on government projects
- Industry levies
- Innovation grants, venture capital and seed funding for startup companies
- Skills development and training policies
- Support for pure and applied research at universities
- Grants to foster greater collaboration between industry and universities
- Fostering international collaborations through trade fairs
- Funding centres of excellence
- Research and business networks and forums
- Funding research directly through government research bodies such as the BRE and CSIRO etc.

The role of unions in driving innovation

Another institution which plays an important role in industry reform is the union movement. While disempowered in many countries over the last twenty years, few would argue that unions have played and continue to play a critically important role in driving innovation through their scrutiny of management practices, particularly in areas such as occupational health, safety and welfare. However, the role of unions in the innovation process has been contentious and opinions are polarized as to whether they are a positive or negative force for innovation. For example, some argue that the relatively difficult and, often, openly aggressive employer/union relations in the Australian Building and Construction Industry have been a major impediment to reform by institutionalizing restrictive trade practices and forcing up wages without commensurate increases in productivity (Garling 1991; RCBCI 2002b; RCBCI 2002c). For example a recent report by the Business Council of Australia (BCI 2012) which is highly critical of unionization in the construction industry claims that Australia has become a high cost, low productivity nation. Productivity has consistently (but slowly) grown in construction over the last twenty years but has not kept pace with wage growth. According to the report, Australia is a great place to invest but not to execute projects (90 per cent of infrastructure projects run over time and budget). Although the BCI report has been criticized for being based on flawed research and for being one-sided (Best 2012), it reflects widespread opinion that unions are an impediment to innovation and resist managers' desire to bring about change at every turn. The main problem appears to emanate from the negative spill-over effects of innovation on traditional trades and industries on which union members jobs depend. For example, a new technology such as off-site fabrication is likely to make certain trade groups redundant, combine several trades into one function or create an entirely new function that needs to be accepted into existing trade power structures. Conversely, others argue that union activities have had a positive impact on industry reform over the years by driving change and industry improvement in areas that affect workers' rights, conditions and health and safety. For example, campaigns to prevent injuries by the unions have resulted in many innovations to reduce the use of hazardous materials such as asbestos, material packaging and handling and falls prevention in the industry. Furthermore, it is widely accepted that the Union Green Bans of the 1970s in Australia were successful in stopping insensitive developments along Sydney's foreshore and in highlighting the need for reform to Australia's heritage laws. More recent innovations which have been driven by union activism in Australia include the introduction of the construction industry long service scheme where developers pay a contribution to long service leave of construction workers on the basis of service to the industry not service to employers (since the casualized nature of the industry exempted the majority of construction workers from the national scheme). The establishment of licensing in the demolition and asbestos removal industry and the introduction of mandatory redundancy trusts and weekly superannuation payments for construction workers are examples of other important social innovations brought about by union agitation. While these may not be seen as innovations by those employers that do not want to pay for them, seen from an independent perspective they are innovations nonetheless. As MacLeod and Clarke's (2010) report to the UK government showed, unions represent an important independent voice for workers and play a significant role in strengthening employee engagement which is directly correlated to levels of company innovation. MacLeod and Clarke's research provides numerous examples to link innovation to union representation. One such example is BAE Systems, which is an international aerospace company of over 32,000 employees. BAE is widely recognized as being at the forefront of innovation in its industry and at the heart of BAE's business is a close working relationship with unions which recognizes and respects the rights of workers to union membership. At the same time, the unions respect the responsibility of management to manage the business. As one BAE Systems manager stated 'This approach has led to a real openness between management and the unions to new ideas and influences that are at the heart of the company's very creativity' (MacLeod and Clarke 2010: 101).

The role of professional institutions in driving innovation

Like the unions, professional institutions have emerged to protect the interests of their professional membership base by defining and defending their roles in the construction process and by setting standards of entry for aspiring new entrants into those professions. Professional institutions have also sought to control domains of knowledge in their areas of practice through ongoing training and development programmes and through the accreditation of education programmes in universities and in industry. These institutions and their members, who teach at accredited universities, guard their knowledge territories vigorously since their future depends on maintaining an enduring professional identity which potential clients can associate with (and pay for).

While on one hand these barriers to entry help to maintain quality of service to clients of the construction industry, they can also represent a significant barrier to innovation. Not only do they act to resist innovation in curriculum content and pedagogy but they also prevent the establishment of a multidisciplinary knowledge-base which we have shown is critical to innovation in the industry. New ways of working are likely to challenge and even destroy traditional core knowledge domains and competencies and professional institutions will need to shift to reflect these profound changes if they are to assist rather than hinder the innovation process.

Cultural impediments to innovation

Dumbelton (1986) and Toole (1998) argue that positive attitudes towards innovation are the most fundamental building block of reform. Yet, in a study of innovation in the construction sector, Loosemore et al. (2002) found that innovation is widely perceived as an academic activity, which in most instances produces theoretical results of little commercial value. There are clearly many reasons for this culture but research indicates that it has its roots in ten main construction industry characteristics (Loosemore et al. 2003; Dainty and Loosemore 2012):

1. The increasingly theoretical research which is undertaken at 'research-intensive' universities (this is discussed in more detail in chapter five)
2. The inability of many academics to communicate the value of their results to industry
3. The masculine culture of the construction industry
4. The lack of diversity in the industry and the under-representation of minority groups
5. The tradition of physical activities which cannot be learnt in a classroom environment
6. The high pressure project environment which provides little time to reflect on the potential value of new knowledge
7. The underlying commercial values of the industry which produce a short-term and narrow focus on financial returns
8. The vertical fragmentation of the industry into a multitude of SMEs which do not have the foresight, inclination or funds to invest in R&D, training and education
9. The horizontal fragmentation of the supply chain into separate design, construction and operational activities which prevents the integration of knowledge and collective responsibility needed to innovate
10. The tradition of passing risks to the lowest common denominator in supply chains and starving them of resources to invest in innovation.

The industry's fragmentation is a particular problem which merits further discussion. For example, Lenard (1996) found a widespread belief amongst contractors that the primary responsibility for both product and process innovation lay with subcontractors and manufacturers respectively. Subcontractors are increasingly expected to take the risks of performance by principal contractors and are perceived to have the expertise and knowledge to make potential improvements in their specialist areas of expertise. There is a perception that principal contractors can only coordinate and manage these innovations to ensure that a project is delivered on time and within budget. The problem with this approach is while subcontractors may well have more connectivity with innovations in their field, most are too concerned with day-to-day survival to innovate in the long-term. High gearing, dependency on regular cash flow and relatively low margins makes long-term investments in new ideas far too risky for most small subcontractors. So the dilemma facing the construction industry is that principal contractors with the resources to innovate are unwilling and technologically less able to do so while the subcontractors who are capable and willing to do so do not have the necessary resources. It would seem that the problem of innovation in the construction sector is primarily one of resource and risk distribution. Clearly, what is needed to break the under-resourcing cycle is a system which enables contractors and subcontractors to spread the risks and benefits of innovation more evenly.

Enablers of innovation in the construction sector

Having discussed the barriers to innovation, recent interviews by Loosemore and Holliday (2012) with leading innovators in the UK construction sector showed that there are four main groups of enablers which could drive greater innovation in the construction sector:

1. Collaboration
2. Regulation
3. Skills, education and research
4. Leadership.

Collaboration

The basis of the collaboration argument is that few firms in the construction industry have the full range of competencies necessary to innovate. So they must collaborate to do so. Collaboration not only creates the broad knowledge and skills-base to generate and convert new ideas into reality but also spreads the significant risk associated with innovation. To drive innovation, collaborative arrangements must extend beyond the traditional boundaries of the construction industry. Traditional knowledge domains within the construction sector are too narrow to understand and manage the enormous scope of impact of the industry's activities on wider society (in social, economic and ecological terms). The poorly understood interdependencies between the industry's activities and the communities in which it does business presents managers with new problems which are not well understood. The solutions to these challenges will also cause us to question the efficacy of vertically hierarchical and integrated business models in favour of horizontally dispersed network models with permeable boundaries. This will require construction business, communities and governments to work in new organizational configurations which are likely to challenge traditional competitive relationships and perceptions. Increasingly, to deal with these new challenges, individuals, firms and governments will need to innovate and it is in new relationships and at the intersection of different knowledge domains that the solutions to many of the industry's future challenges will lie. The most enlightened organizations will realize this and build this into their business structures and strategies. To them, notions of interdisciplinarity and collaboration will replace traditional competition as a way to share complementary ideas, resources and capabilities to mutual advantage. New communication technologies can help facilitate this process by enabling firms to work in a more integrated and collaborative fashion. Procurement reform will also be critical to this process. Partnerships, alliances and relational contracting will become particularly important mechanisms to enable more integrated working. These new approaches will need to enable 'deeply integrated' supply chains where suppliers, manufacturers, designers, contractors, subcontractors, operators and clients work collectively to share the risks and rewards associated with innovation in an open, transparent and collaborative environment of collective responsibility where personal interests are put aside in the interests of the whole and where resources are shared and used where they best fit. Deeply integrated supply chains create a single point of responsibility for innovation, improve communication and allow the early involvement of firms in the development process preventing solutions being locked-in before all available knowledge has been brought to bear on them.

Regulation and certification

The basis of the regulation argument is that much of the construction industry will not innovate voluntarily. Construction is a compliance-based sector with a long-tail of low performing firms which must be encouraged to change through regulation and legislation which mandate minimum standards and create incentives to innovate. The industry also needs standard methods and metrics for measuring and demonstrating the benefits of innovation. These targets need to be incrementally improved over time to drive up performance. However, the need for regulation must be balanced with the dangers of over-regulation which when badly designed is also known to stifle innovation through increasing red tape and compliance costs which can slow down the innovation process and make it economically unviable.

Skills, education and research

Innovation requires both a deep and a broad gene pool, which provides the expertise and diversity which is the basis of innovation. Attracting the best people (deepening the gene pool) is not an easy task given the increasing competition from other industries for human resources. This will depend on improving the industry's image and reputation by improving working conditions and performance and by highlighting the important and positive contribution it makes to the economy, culture, environment and wider society. Broadening the gene pool will require the industry to address the under-representation of minority groups at all levels. For example, women are seriously underrepresented at senior levels in the industry, minority groups such as migrants are often poorly treated and there are few education and career pathways for trade operatives to work their way up into management roles. To alleviate this problem, equity and diversity initiatives need to be taken seriously and education needs to be more flexible in enabling people to learn at work, to recognize prior and experiential learning on-the-job and to offer clearer pathways and transitions between all levels of education in the industry. It is also important to ensure that education is relevant to what industry needs and that it contributes to the performance of firms in the industry, while at the same time providing the broad-based liberal education needed to think differently. Finally, there needs to be stronger ties between universities, governments and commerce to ensure the industry can independently monitor the benefits of innovation and learn lessons to improve future processes.

Leadership

Four sources of leadership are important to innovation in construction:

1. Major clients (private and government) must lead by demanding more innovation and creating a competitive environment where innovation can flourish. Instead of focusing on price alone (which often disadvantages innovative firms), they should recognize the 'value' delivered by innovation over the life-cycle of their investment. They should also enable supply chains to have a stake and say in the way it is designed, constructed and performs.
2. Major firms in the construction industry have a responsibility to lead by driving innovation through their supply and demand chains. They can do this by: investing in R&D; educating clients about the benefits of innovation; being more willing to support new ideas; helping smaller business partners to do so; sharing risk and reward; and continuously striving to improve performance on their projects.
3. Innovation is more likely to be driven by the adoption of innovations from outside the industry than from within it. So the leadership of manufacturers and service providers have a critically important role to play in collaborating to drive innovation within the industry.
4. At an industry level there is a need for a champion to catalyse innovation in the same way that Latham did in the 1980s and Egan did in the 1990s in the UK. These people caught the imagination of the industry and engendered a universal appetite for reform never seen in the UK construction industry before.

Conclusion

The aim of this chapter was to discuss the barriers and drivers to innovation in the construction sector. While the sector is often derided for being 'low innovation' the reality is that it is a highly creative industry that is adept at solving complex operational problems on a project-by-project basis. Many of these creative activities are not detected by traditional measures of R&D and the challenge that faces the industry is how to appropriate the benefits of this creativity across projects and throughout the supply chain. Bringing the discussions together we can conclude that innovation could be increased by:

- stronger thought leadership from clients and major firms in the industry
- closer relationships between innovators, clients and end-users
- shorter and more integrated supply chains
- reduced emphasis upon price and increased the focus on value
- developing metrics to measure innovation.
- developing methods to value innovation
- collaborative procurement systems which emphasize negotiation not competition
- more equitable sharing of risk and opportunity associated with innovation
- more realistic resourcing of projects
- a willingness to work with competitors for mutual benefit
- improved communication between research, production and marketing functions
- imaginative and effective training to develop people's innovative capabilities
- a more diverse workforce
- systems to measure and reward innovation
- an understanding of market forces and of future trends
- an educated and courageous client-base
- continuous, trusting and confident relationships with clients and business partners.

It is clear that the adoption of many of the above recommendations will require a fundamental mind-set shift within the construction industry. While there are clearly many institutional, organizational and cultural barriers which can prevent this from happening, the upside is that those that address them have the potential to reap extraordinary rewards among competitors who remain unsophisticated and slow to adapt to an increasingly uncertain and dynamic business environment which is full of new opportunities.

In the next chapter we move outside the immediate confines of the project environment into the broader realms of management research to provide more strategic insights into the ways firms in the construction industry can respond effectively to this challenge.

4 The history and future of innovation in the construction industry

> Those who cannot remember the past are condemned to repeat it.
> George Santayana, Spanish American philosopher, essayist, poet, and novelist

Introduction

As Hamel (2002: 10) points out, every age brings its own unique blend of promise and peril. And it is those who have been able to 'imagine a new reality' by escaping the linear extrapolation of their past that tend to succeed. While many would argue that many contemporary challenges like climate change have no parallels in the past and that 'looking in the rear-view mirror' is no longer a reliable indicator of the future, the quotation by Santayana was chosen to begin this chapter because it illustrates the importance of history in understanding the future. As Professor Geoffrey Blainey AC, one of Australia's most pre-eminent social and economic historians recently said, 'while it may be hard to learn from history . . . it's even harder to learn if you know no history' (Oldfield 2012). However, history shows us that humans are constantly caught out by the future. Predictions of the paperless office, personal jetpacks and holidays to the moon are just a few examples of many wide-eyed predictions which have proved to be completely wrong. Even the greatest thinkers of our time have been unable to predict it with any degree of accuracy. For example, John Maynard Keynes wondered how people in the twenty-first century would fill the plentiful leisure time that new technologies would afford them. Aldous Huxley's *Brave New World* imagined a world where people would need to take a drug called Soma to alleviate the boredom this would bring. As most of us unfortunately know, the reality has turned out to be somewhat different and Aldous Huxley's drugs are more likely to be prescribed to alleviate the stress of overwork than to alleviate the effects of boredom. According to Rosen (2010), the industrial revolution fundamentally changed our perception of time, moving society from an 'organic' pre-industrial version of time which was based on seasonal rhythms to 'clock time' as measured by the artificial dictates of mechanical instruments. More recently, the electronic revolution has moved societies to the hyper artificial measurement of 'digital time'. New communication technologies such as the internet and mobile phone have created a twenty-four-hour working day where many of us appear to have too few hours to meet the demands made upon us – a phenomenon called 'time famine'. In his recent book *Thinking Fast and Thinking Slow*, Nobel Prize winning economist, Daniel Kahneman warns of the consequences of this trend for innovation (Kahneman 2011). According to Kahneman, modern digital communications have increased our expectations about the speed at which we receive information, often at the cost of the quality of information and insight. Consequently, he argues that almost everything we read about in the media is useless in predicting the future. This is not because experts do not know their field but because they are pandering to a public who are generally not interested in anything but a shallow 'sound bite' as to what the future may hold.

While our many attempts to predict the future have generally failed, it doesn't make crystal-ball gazing worthless. History also tells us that it is the firms which attempt to anticipate, understand and organize for the future which will succeed, no matter how far from the truth their predictions were. Thinking about the future and what it may hold has many advantages. Not only does it position firms to capitalize on future potential opportunities, even if they may be different to those anticipated, but it also minimizes the chances of falling victim to the potentially devastating effects of 'Black Swan' events which often bring seemingly healthy companies down. The idea of Black Swan events was theorized by Nassim Nicholas Taleb in his bestselling book *Fooled by Randomness: The Hidden Role of Chance in Life and in the Markets* (Taleb 2001) and then elaborated in his following book *The Black Swan: The Impact of the Highly Improbable* (Taleb 2007). The Black Swan is a metaphor that describes an event that is a surprise (to the observer), has a major effect and after the fact is often inappropriately rationalized with the benefit of hindsight. In English, when the phrase was coined, all swans were thought to be white, and its subsequent discovery illustrated the fragility of any system of thought, of the phrase's underlying logic, as well as any reasoning that followed from that underlying logic. Taleb argued that it is these highly improbable, high impact 'Black Swan' events that often change our lives. These 'outlier' events are hard to predict and are beyond the realm of normal expectations in history, science, finance and technology. The probability of such events is therefore very hard to compute using scientific methods (owing to the very nature of small probabilities) and there are many psychological biases that make people individually and collectively blind to the massive role they play in historical affairs. Taleb regards almost all major scientific discoveries, historical events, and artistic accomplishments as 'black swans' – undirected and unpredicted. World War I, the uptake of the internet, the dominance of personal computers, World War II, the September 11 attacks and the global financial crises are all examples of such events.

With the above in mind, the aim of this chapter is to provide a historical overview of how firms in the construction industry have adapted over time to their changing business environment. We also discuss the main periods and types of reform experienced by the construction sector since the Second World War and place the organizational responses of firms in the sector within context of strategic management thinking at the time. The chapter concludes by taking its own risk in setting out possible future trends which may affect the construction sector into the future. And we speculate as to how the industry might adapt within the context of recent developments in strategic thinking.

The post-war years

The Second World War left Europe with a bewildering rebuilding challenge. Europe's population was without basic accommodation and essential services and massive investment programmes were initiated to rebuild the residential, commercial, industrial, transport and civil infrastructure which had been destroyed. In the Soviet Union alone, 25 million people had lost their homes, in Germany 2.4 million homes were needed and in Yugoslavia, Poland and Greece over 20 per cent of the housing stock had been destroyed (Sebestye'n 1998). This led to an unprecedented building boom across Europe with as much as 10 per cent of world GDP being invested in the construction and engineering industries. Fuelled by increasing urbanization (fourteen new towns were approved between 1946 and 1950 in the UK alone), the construction industry across Europe grew rapidly in response to the insatiable demand. The vast majority of the construction industry at this time was traditionally craft-based and dominated by family-owned, small-to-medium-sized businesses serving localized community needs and employing large directly employed workforces. Since the structure of these businesses was not geared to deliver the volume and speed of production required by the unprecedented demand for new buildings, firms began to innovate with new forms of construction methods and organization, often borrowed from manufacturing. For example, many firms invested heavily in mechanization and off-site prefabrication and experimented with new materials such as GRP and other industrialized systems such as new in situ processes. They also experimented with Taylor-inspired methods such as dimensional coordination and modular construction, and many companies grew rapidly by acquiring manufacturing and supply companies in an attempt to thwart dwindling raw material supplies, speed up the production process and deliver cost advantages over their numerous competitors. As the need for supporting social and economic infrastructure emerged, many general contractors rapidly grew into the industry's first truly national and international firms. These firms eventually diversified into functional divisions and learnt to use new technologies and engineering solutions developed during the war to build increasingly sophisticated structures, bridges and services. Many firms also diversified overseas and opened major regional offices, although in general, construction firms remained smaller in average size than the multinational manufacturing firms that had grown to meet the massive post-war demand for products. Large international US firms like Bechtel, Parsons and Morrison-Knudsen became the world's leading contractors in engineering while UK firms like Laing, Taylor Woodrow and Wimpey grew out of the housing sector to dominate general contracting and infrastructure in many overseas markets. During the 1950s several Korean, Japanese and Eastern-Block companies also appeared as competitors on the global construction market.

1960s – the decade of growth

During the stable and prosperous 1960s, firms in the construction and engineering sector continued to diversify, expand and acquire competitors. This led to large multinational conglomerates emerging which effectively acted as mini capital markets. The rationale was that large firms benefited from economies of scale and a lower cost of capital, and attracted more talented human resources than smaller firms. This development reflected a trend in strategic management thinking at the time which was popularizing formalized long-range planning based on a functional view of the firm. This trend also emphasized the benefits of centralized marketing, human resources, finance and production functions and was influenced heavily by the formative work of Chandler (1962) and Ansoff (1965) who, as we pointed out in chapter one, promoted the view that organizations were a triangle with a few elite strategists at the top who were presumed to have all the information needed to predict and plan for the future in great detail. According to Langford and Male (2001) this thinking was embraced by the construction sector which continued to experience stability and high growth rates stemming from residual post-war investments in housing and infrastructure. Since markets were not changing and the business environment was still stable, strategy in the construction sector had an internal focus on systems, procedures and organizational structures that could provide competitive advantage, mainly in cost differentials. Competition was also relatively low because of the abundance of work and because of the market specialization that had occurred within firms organically from established relationships in specific sectors.

Construction industry reform

It was during this period that concerns with the construction industry's performance first began to emerge with the UK government commissioning a number of prominent investigations into the construction sector's efficiency. Murray and Langford (2003) provide an excellent review of the main government reports during this period and up to 1998. However, the most influential of these during the 1960s were Emmerson (1962), Banwell (1964), Lofthouse (1965) and Higgon and Jessop (1965) which all criticized the construction industry for its resistance to adopt new technologies and modern management methods. These reports also exposed high variability in the industry's performance and recommended closer integration of design and construction processes and greater industry unity to collectively change attitudes towards innovation. Critical to this was improved communication within the industry, the rationalization and speeding-up of payment procedures (particularly to the growing numbers of subcontractors) to reduce under-capitalization of the industry and the greater use of fixed price contracts underpinned by longer-term wage settlements. A lack of investment in training was also identified as an important issue and it was argued that a workforce of well-qualified site agents, foremen and project managers was needed with skills to enable the more efficient use of labour, better use of machinery and more economic use of materials. As well as a comprehensive review of UK building standards, there were also calls for common education and training across professions and recognized standards of behaviour and workmanship across the industry. Furthermore, the industry was lamented for its poor working conditions and its poor public image. Subcontracting had also become more common during this period in response to greater building complexity and specialization and there were associated concerns about poor quality control, disputation and industry fragmentation which led to calls for a return to more direct forms of employed labour.

Price collusion amongst an ever smaller number of large contractors was also an issue and arguments were put to encourage more stringent pre-qualification and selective tendering from approved lists of contractors. One strategy put forward was the creation of a central body for the exchange and provision of information to allow contractors to benchmark their performance. Another solution was the adoption of two-stage tendering where firms were first competitively selected on management capability and then on the basis of price as designs developed. To address the growing adversarial culture in the industry, better up-front planning of projects to reduce unnecessary variations was proposed as was a standard form of construction contract and reduced retention requirements for subcontractors. Finally, the public sector was asked to take a lead and show a greater willingness to be innovative with procurement methods.

1970s – the decade of shock and reflection

In contrast to the stable 1960s, the 1970s were characterized by stagnation and inflation. The oil crisis of 1973–1974 decimated the construction industry, reducing domestic and international demand for construction and producing more conservative corporate strategies with a focus on home markets, core businesses, niche markets and systems of financial control.

Construction industry development

As Langford and Male (2001) point out, during this period of global economic turmoil, construction firms were faced with three strategic alternatives: to shrink (retrenchment); to specialize; increase their internal efficiency and expand into existing markets (expansion); or to seek out new markets (diversification). Specialization turned out to be the most effective strategy to survive the recession as construction firms were re-organized from functional units into strategic business units to serve specific customer and competitor profiles. Falling domestic demand for construction and infrastructure also forced contractors to explore new ways to compete. During this time, many construction firms pursued mergers and acquisition and borrowed heavily to explore new markets at home and overseas. And in response to this diversification strategy, portfolio management techniques such as the Boston Consulting Group's Matrix developed to help firms manage the inter-relationships between different parts of these increasingly dispersed business empires.

Construction management research

During this period, the construction management research community began to evolve in the UK underpinned by special government funding programmes to encourage greater research and development in the industry. Research in construction management was very much in its infancy with very little written about strategy per se. Early researchers such as Newcombe (1976) compared procurement processes employed in the construction industry to those commonly found in manufacturing and laid the conceptual foundations of the field. This enabled construction management scholars to start experimenting with theories and ideas from mainstream management to better understand strategic decision-making in construction firms. Lansley et al.'s (1979) work was also formative during this period. They found that construction firms were best able to adapt to their environments when they had a clearly defined vision for the future and when managers were perceived to be dedicated to those objectives. They also found that successful construction firms had an internal consistency between managers' behaviour and espoused values, high levels of staff morale and satisfaction, effective intelligence gathering about their business environment and a concern for staff welfare and feedback. Successful firms at the time also had effective corporate planning systems in place and were sensitive to external competitive positioning rather than simply internal efficiency.

1980s – the decade of change and reflection

The 1980s marked the beginning of business globalization as international barriers to trade in products and services fell. The emergence of Japan as a new economic force raised increasing questions over the efficiency of traditional production processes used in the West. In response managers became focused on corporate restructuring, divestment, downsizing and the pursuit of efficiency through better strategic planning, the elimination of waste and better quality control. Managers also focused on addressing the many shortcomings of earlier management strategies which had caused firms to diversify into multiple businesses. New research during this period showed that random diversification into unrelated businesses was rarely successful and that even if a firm's individual businesses were successful it may not translate into success at an aggregate corporate level (Prahalad and Bettis 1986). To be successful, diversified corporations had to seek and exploit synergies between strategic business units to justify their existence.

Henry Minztberg's (1984) work was also highly influential. He attempted to synthesize strategy into five different approaches:

1. A conscious rational future plan
2. A tactical ploy to out manoeuvre competitors
3. A historical pattern which evolves organically over time
4. A positioning process to fit the firm to its environment or
5. A social perspective ingrained informally in the collective understanding and culture of an organization.

Michael Porter's work (1980, 1985, 1987) was also formative during this period as strategic thinking focused on the concepts of competitive advantage, differentiation and the strategic positioning of firms in relation to their competitors. It was argued that competitive advantage was achieved by matching a firm's strengths and weaknesses with the risks and opportunities in its business environment. However, late in the 1980s, the limitations of Porter's approaches began to emerge leading to alternative 'resource-based' views of strategy which emphasized firm-specific 'distinctive' competencies and resources as the key determinants of competitive advantage (Prahalad and Hamel 1990).

Construction industry development

During the 1980s changes to the traditional craft-based nature of construction continued with the increasing use of technology and off-site prefabrication – forcing the integration of design, manufacturing and construction. Architects also began to lose their traditional leadership role as both designers and managers of the building process, particularly on the more complex building projects where specialist project management consultants were starting to emerge. Architectural firms responded through diversification, acquisition and mergers – forming large integrated multi-disciplinary businesses which offered both design and project management services. Construction firms which had remained largely homogenous until this point also began differentiating into two broad categories of firm: contractors which provided project management services and those providing traditional contracting services. The transition of larger contractors into project management service providers was a logical response to an increasingly risky construction market. This enabled them to pass most of their risk to the many smaller and more specialist subcontracting businesses that had moved into the production gap. Another risk mitigation strategy employed by larger construction firms during this period was to move up the supply chain (forward integration) into property management or down the supply chain (backward integration) into product manufacture. Many construction firms also borrowed money to pursue growth, diversifying into areas such as building materials, coal and mining, plant and waste disposal. The main reason behind this diversification strategy was to reduce exposure to economic cycles in the construction market, to grow companies into less risky but more profitable areas and to seek potentially profitable synergies between different businesses. However, in time, many of these newly diversified business groups failed due to a poor understanding of the risks in these new markets. Furthermore, many of these acquired businesses were cash-hungry resulting in unbalanced business portfolios which drew resources away from core business activities which were efficient and profitable.

There were also important changes in company governance during the 1980s in the construction sector. Company boards of directors were reduced in size and many firms established executive committees to provide specialist advice in different business activities. Furthermore, the role of independent directors became more important in providing advice to assist executive teams with making strategic decisions. As Langford and Male (2001) point out, construction firms had not invested in management development during the 1970s so many managers had to be imported from other industries. Increasing industrial unrest and low profitability in the sector also forced many firms to leave or to diversify into new unrelated markets. Inspired by the notion of the 'flexible firm' (Atkinson 1984) the construction industry also experienced one of its most profound changes in industry structure – the emergence of subcontracting as the dominant form of organizing. Subcontracting had existed long before the 1980s but grew rapidly during this period in response to increasingly unpredictable demand, increasing building complexity, increasing pressures to reduce fixed costs and growing client demands for greater price certainty (by securing fixed price work packages). Subcontracting was also encouraged during the late 1980s by political ideologies and government policies which were designed to break the unions which were seen as a barrier to innovation and productivity. During this period of dramatic change in UK politics, small businesses were seen as the key to innovation.

Construction industry reform

However, towards the end of the 1980s the problems associated with subcontracting were beginning to emerge. Reports such as NEDO (1983) and NEDO (1988) exposed the problems of fragmentation, quality, training, loyalty and conflict that the subcontracting model had induced. Data in the UK showed that only 1 in 5 buildings were completed in an efficient time and the ground was laid for a profound and long-lasting industry reform agenda which continues to this day. The NEDO reports cast light on the inefficient work practices of the UK construction industry which had cemented into place restrictive procedures and practices. These had protected the inefficient and held back external investment in the industry. The industry was also found to be highly confrontational in its practices and culture which was leading to many costly disputes on projects. NEDO's recommendations were that the building and construction industry could build more quickly and efficiently if customers played an increased role in the procurement process by pushing for faster times. It also recommended that clients be clearer about their objectives up front during the briefing process so that changes in scope could be avoided and that procurement systems with a single point of management responsibility be adopted to improve accountability and simplify management oversight. The use of professional project managers to manage the construction process from inception to completion was also encouraged and other measures to improve performance included better design management, earlier integration and reduced reliance on subcontractors and more use of advanced technologies to speed up construction. The NEDO reports also found that systems in the industry were unable to accommodate client changes without a disproportionate impact upon performance and that management skills across the industry were generally poor. Furthermore, the large number of separate specialists involved in the design and construction process created too many interfaces to manage and ambiguities over risk distribution in contracts created a confrontational culture that was detrimental to project performance. Supervision and quality control of subcontractors was also found to be inadequate.

Between the two NEDO reports, the UK's Building Employers Confederation launched its own investigation of construction performance in 1985 (BEC 1985). This was designed to specifically address increasing concerns about the relatively low productivity, cost effectiveness, speed and quality of the UK building industry, compared to the US. The study concluded that UK building projects were significantly slower, more complex, less standardized and more difficult to build than US buildings. The UK also used significantly less technology in the construction process and fewer prefabricated, mass-produced and standardized off-the-shelf components than US buildings. Furthermore, while in the US designs were normally 100 per cent complete before site work started, UK contractors had to commence work with incomplete designs leading to significant rework, waste and dispute.

During the 1980s, major industry clients were also becoming increasingly frustrated that building was a confrontational experience, cost too much, took too long and did not always produce reliable results. In a move which was to mark the beginning of greater client leadership of the industry, the British Property Federation produced a major report and a new approach to procurement which encapsulated many recommendations of previous reform reports into the sector (BPF 1984). The BPF was a relatively small but influential group of clients which highlighted some important issues and created some lively debate. The aim of the BPF system, which was unashamedly client-driven, was to promote many ideas from the US where costs and production times were far lower than in the UK. It also sought to promote a more efficient and cooperative method of organising the whole building process which eliminated practices that absorbed unnecessary effort and time and obstructed progress towards completion.

In other parts of the world like Australia, construction sector reform was also taking place, although with a very different focus on industrial relations. Here, the Australian Procurement and Construction Council (APCC 1997), the Australian Federation of Construction Contractors (AFCC 1988), The Maddock Report (Maddock 1989), The Department of Industry, Technology and Commerce Report (DITAC 1989) and The Review of Structural Efficiency Principle Report (DIR 1989) sought to address the relatively high level of claims and disputes in the construction industry and the low predictability of project performance. Collectively, these reports established the need for, and direction of, labour market reform as a catalyst for productivity and efficiency improvements in the Australian building and construction industry. This resulted in a radical overhaul of Australia's labour award systems and the first moves towards a system of decentralized enterprise bargaining. The philosophy underpinning these reports was that a more flexible and highly skilled labour force was required to improve the efficiency and competitiveness of the industry and to provide workers with access to more varied, safer, secure, fulfilling and rewarding jobs (de Valence 1997). Prior to this, the Commonwealth and State Governments had become increasingly concerned about industry malpractices and in response, had deregistered the Building Labourer's Federation (BLF) by legislation in Commonwealth industrial relations' jurisdictions and territories. The Commonwealth government had also introduced a Code of Conduct for all contractors and developers of Commonwealth construction projects (ICE 1998).

Construction management research

During the 1980s our understanding of construction management as a distinct field of study also grew rapidly. In particular, researchers like Lansley (1981) Hillebrandt (1984), Seymour (1987) and Hillebrandt and Cannon (1989) began to reveal the nature of innovation in the construction industry and how it was restricted by the fragmented nature of design, construction, production and manufacturing. The growing body of research indicated that skilled labour was responsible for much of the innovation in the construction industry and that innovation was also highly dependent on the quality and longevity of relationships with business partners, subcontractors and suppliers. Innovation in construction was often inhibited because most competitors tended to have the same relationships. Furthermore, workers at operational level rarely saw the benefits of their ideas. Most innovation within the industry appeared to occur in the 'process' of managing resources more efficiently. 'Product' innovation tended to occur outside the industry within manufacturing firms. However, the balance between process and product innovation in construction varied with economic cycles – product innovation being more likely in times of recession when significant shifts in competitive advantage were sought and process innovation in times of boom when incremental shifts were sought.

For the first time we also began to understand that demand in the construction industry was client-generated and that there were many different types of clients (experienced, inexperienced, regular, intermittent, one-off etc.)

which required different procurement strategies. For example, inexperienced clients dealing with simple buildings were better served by traditional methods of procurement which separated construction and design. In contrast, more experienced clients with sophisticated needs were better served by management contracting strategies and integrated forms of procurement which facilitated more direct involvement in the building process. It was also found that economies of scale were not particularly easy to gain in construction because of the project-based nature of production, the fragmented nature of supply chains and the relative uniqueness of each project. However, economies of scale could be achieved by integrating the supply chain and production process, focusing on repetitive work, specializing in one type of product, service or market, seeking repeat work with limited numbers of clients or by diversification into finance, design and manufacturing to maximize control over the entire supply chain.

It was also argued that every construction firm had a unique competitive and operational environment determined by their strategic choices in relation to markets and business partners. Understanding this environment and the market a firm chose to operate within was the key to survival. To this end, two distinct markets were identified in construction with different economic forces at work. First there was the contracting market where companies tendered for projects in response to client demand. Success in this market was found to depend on client relationships, reputational, technical, management and contractual skills. Second there was the speculative market where success depended more on analytical, financial and entrepreneurial skills.

1990s – the decade of reform

During the 1990s, accelerations in technological advances made buildings increasingly complex. Growing costs pressures, declining union power and the need to reduce overheads and to de-risk led to further specialization and fragmentation of the construction industry. Dramatic economic and political changes occurred around the world. The end of the Cold War and the collapse of the Soviet Union, the growth of the EU and the demise of Japanese industry led to the growth of corporate networks and alliances, joint ventures and continued restructuring towards service-based organizations. Many firms also moved production offshore to exploit cheaper resources.

Many of these changes reflected thinking in strategic management at the time which promoted the idea of core competencies (Prahalad and Hamel 1990). Management thought during the 1990s also highlighted the importance of innovation to sustained competitive advantage and revealed that most firms innovated incrementally through minor but continuous and often unnoticeable refinements to their existing products and services rather than through radical new innovations (Kay 1993). This research led to deeper understandings of the critical resources a firm needed to make it grow. In particular there was a new focus on the role of intangible assets (such as reputation, organizational culture, tacit knowledge and social capital) in securing competitive advantage. Nonaka's

(1991) work on the knowledge creating company was especially influential. He argued that knowledge creation was the basis of innovation but that few managers at the time understood the nature of knowledge and how to manage it. He criticized the Taylorist traditions of Western management which proposed that the only useful type of knowledge for business decision-making was formal, numerical, systematic and scientific (explicit knowledge). In contrast, Nonaka argued that innovative and highly responsive companies like Honda, Canon, NEC and Sharp etc. had found another way to think about knowledge. The centrepiece of this Japanese approach was that management was not simply a process of analysing objective numerical statistics but was also dependent on tapping into the tacit and often highly subjective insights of individual employees. This in turn depended on their personal commitment and sense of identity with the business. Nonaka also argued that in innovative companies, innovation was not a specialized activity (the province of the R&D or marketing department), but something everyone was involved in. New knowledge, he argued, depended on redundancy which gave people space for dialogue and open communication and created a common 'cognitive ground' for people to exchange tacit knowledge (something which Taylorists sought to eliminate). He also argued that innovations always began when an individual had an idea. For example, a researcher discovers something new, a manager senses a new market trend or a shop floor worker comes up with a new improvement to work processes. This tacit knowledge is then transformed into explicit organizational knowledge through the use of metaphors, analogies and models which have direct implications for how an organization designs its structure and defines its managerial roles and responsibilities.

During the late 1990s, the concept of 'dynamic capabilities' also emerged. This referred to the capacity of successful firms to adapt their resources and capabilities to respond to changing environmental circumstances and to exploit new opportunities (Teece et al. 1997). The ability to do this was found to be influenced by path dependencies (established routines, history and previous investments in established routines which inhibit learning of new routines) and a firm's current asset position (the ability of a firm's structure, technologies, people, finances, reputation etc. to change). This work refined the 1980s' resource-based view of the firm which had argued that it was a firm's 'non-substitutable' and 'imitable' resources which were its main source of competitive advantage. These resources were physical, human, economic, intellectual and social. Michael Porter's (1998) work on competitive advantage also remained influential and developed into a more balanced view of strategy, retaining the notion of strategy as 'fit' but giving greater credence to 'content' of strategy rather than just the 'process' by which it was developed.

Construction industry development

These changes created a construction sector which was divided into two distinct categories of firms. First there were the local traditional labour intensive, craft-based and low technology firms which undertook repairs, maintenance, alterations, rehabilitation and small-to-medium-sized new building and civil engineering projects in local communities. The second part of the industry was dominated by major national and international firms which had moved up and down the value chain by diversifying into property development, financing and the manufacturing of building materials and which undertook large-scale projects using sophisticated technologies, primarily for government clients. In Western Europe ten firms dominated this market which represented about 30–40 per cent of gross construction industry turnover but only employed about 3 per cent of the total construction workforce (Sebestye'n 1998). These firms included Bouygues, SGE, Eiffage, GTM-Entrepose, Bilfinger and Berger, Philipp Holzmann, BICC, Trafalgar House, Hochtief and FCC. In the UK the equivalent firms were Trafalgar House, BICC, Tarmac, AMEC, Wimpey, P&O, Laing, Mowlem and Costain. And in Japan the industry was dominated by the big five contractors: Kajima, Obayashi, Taisei, Takenaka and Shimizu, each having its own research institute to develop new innovative products, technologies and services.

During the 1990s, over-stretched firms who had borrowed heavily to fund expansion during the 1980s suffered heavy losses and a period of consolidation occurred as smaller firms were bought out by larger firms, sometimes for a token amount. Those firms that survived the predators sought to reduce their vulnerability by drastically reducing gearing, selling assets or disbanding unprofitable business units not seen as core business. They also sought to raise capital through rights issues and to reduce costs through systems improvements, reorganization, salary reductions and large-scale redundancies. The laying-off of many people led to permanent changes in employment patterns within the construction industry with a wider use of casual and agency staff on a project-by-project basis.

During the 1980s diversification period, firms had also developed highly fragmented structures with functional and regional divisions almost operating independently of the central business. Many of these divisions had developed different processes, systems and cultures and often competed with the interests of the main parent group. In response the 1990s saw a refocus on core businesses, the centralization of common areas such as finance and a reduction in the number and size of regional offices.

Construction industry reform

The 1990s saw a raft of new reports into the performance of the construction sector around the world. For example, in the UK, the Centre for Strategic Studies and the National Contractors Group published *Investing in Building 2001* (CSS 1990). In this report, the construction industry's consistently patchy performance in relation to other UK industries was put down to unstructured changes over previous years, its short-term focus, a poor public image, its process orientation, its fragmented organization and educational system, its adversarial culture and its low technology base. The report concluded that leadership in reform of the industry should move to the 'prime movers' who had a moral responsibility to mentor and nurture the industry towards improved efficiency. The recurring themes in this report were responsibility and self-regulation. That is, firms and individuals within the industry should accept greater responsibility for their own competence and the quality of work. Holistic thinking about life-cycle building performance, interdisciplinary working, flexibility and adaptability were also major themes. The report also argued that building users should be more closely involved in the procurement process and that the process of briefing, design, management, manufacturing, construction, commissioning and maintenance needed to merge and extend to include post-occupancy evaluation and feedback into the briefing process. Technology also had the potential to reduce the dangerous, arduous and tedious nature of many building tasks and enable more energy to be devoted to creative thinking and innovation. This could only be achieved by a positive culture and a well-motivated workforce which was driven by leaders who gave meaning to work and who rewarded the dedication needed to produce consistently high quality outputs.

Following the CCS report, in 1993 the *Joint Review of Procurement and Contractual Arrangements in the United Kingdom Construction Industry* was announced to the House of Commons. Chaired by Sir Michael Latham, its terms of reference were to make recommendations regarding reform to reduce conflict and litigation in the construction industry, thereby improving its productivity and competitiveness. An interim report *Trust and Money* was published in December 1993 with the better known final report *Constructing the Team* published the following year (Latham 1994). In this highly influential report (widely referred to as the Latham Report), to which most UK construction reform is often mistakenly traced, Latham concluded that the industry's traditional methods of procurement and contract management created an adversarial culture which caused inefficiencies and ineffectiveness. Latham claimed that addressing these issues could reduce construction costs by 30 per cent over five years. Key recommendations included government leadership to drive reform, legislative changes to simplify dispute resolution and the establishment of a single organization to bring together all sections of the industry and clients (this recommendation resulted in the establishment of the Construction Industry Board in 1995 which was set up to implement, monitor and review the report). The Latham Report also recommended the publication of a wide variety of guidance, checklists and best practice advice in various aspects of the procurement, design and construction processes (later implemented by the Construction Industry Board). It also recommended the establishment of a single central public sector register of consultants and contractors (this recommendation resulted in the establishment of Constructionline in 1998 – a central qualification database of contractors and consultants run as a public/private partnership). Finally, among other things, it recommended the need for more standardized forms of contract, which addressed issues of clarity, fairness, roles and responsibilities, allocation of risks, dispute resolution and payment. Latham drew attention to Australia's main reform body, the Construction Industry Development Association (CIDA) which worked between 1991 and 1995 to create a number of Codes of Practice for the construction sector – including pre-qualification criteria and model projects to trial, monitor and publicize new approaches. The central thrust of Latham's directives was similar to that of the Gyles Royal Commission in Australia (RCBI 1992), which argued that construction clients must use their purchasing power to force industry reform. This led to the establishment of the Construction Clients Forum as a powerful group which was responsible for over 80 per cent of UK's construction expenditure. Despite many of its recommendations not being taken up, the Latham Report did have an important psychological effect by catalysing real collective momentum for change within the industry.

Other important reports into the UK construction industry during this intense period of reform included the 1995 Levene efficiency scrutiny into construction procurement by government (EUCO 1995). This concluded that government bodies were partly to blame for the poor performance of the industry. The scrutiny made a number of recommendations including: better communication within the construction industry to reduce conflict; adoption of a more commercial approach in government; negotiation of deals justified on value for money grounds; and increased training of civil servants on procurement and risk management.

In 1997 the Construction (Design and Management) Regulations also came into force setting out a new framework for addressing the persistently poor health and safety performance of the industry. The regulations applied to everyone in the construction supply chain (clients, architects, engineers, surveyors, designers and contractors), and for the first time highlighted the collective responsibility of the industry to improve its safety performance. The Government Construction Client Panel was also established in 1997 to improve government client performance and to provide a single collective voice for government construction clients on cross-departmental aspects of procurement. And in 1998 the most significant piece of legislation to come out of the Latham Report became law. This was the Housing Grants, Construction and Regeneration Act 1996, better known as the Construction Act which introduced industry-wide adjudication of disputes. In an attempt to speed up dispute resolution, the act required all construction contracts to include provisions for adjudication and also contained new rules to make payment fairer, including a ban on the use of iniquitous pay-when-paid clauses.

The year 1998 was momentous not only for the introduction of the Construction Act but also for the establishment of the Construction Task Force headed by Sir John Egan. His task was to provide advice, from the client's perspective, on the opportunities to improve the efficiency and quality of delivery of construction, to reinforce the impetus for change and to make industry more responsive to customer needs. The Task Force produced the report *Rethinking Construction* (Egan 1998) which emphasized the need for a change of 'style, culture and process' in the construction industry. Five drivers of change were identified: committed leadership; focus on the customer; integration of process and team around the project; a quality driven agenda; and commitment to people. According to the Egan Report, four key processes needed to be improved: partnering by developing long-term relationships based on continuous improvement with a supply chain; developing a sustained programme of improvement for the production and delivery of components; integrating and focusing on the construction process in meeting the needs of the end user; and the elimination of waste. The report set seven annual targets in improving construction project performance relating to costs (10 per cent reduction), time (10 per cent reduction), defects (20 per cent reduction), accidents (20 per cent reduction), predictability (10 per cent improvement), productivity (10 per cent improvement) and profits (10 per cent improvement). To drive these changes a number of streams of activity emerged which included: the Movement for Innovation (which focused on the general construction industry); the Housing Forum (which concentrated on the public and private housing sector); the Local Government Task Force (which promoted the Rethinking Construction agenda within local authorities); the Respect for People Steering Group (which developed a series of toolkits to help improve recruitment, retention and health and safety); and the Construction Best Practice Programme (CBPP) (which was the main dissemination arm for Rethinking Construction). The intention of the Movement for Innovation (M4i) was to bring together and facilitate the exchange of knowledge between those who were committed to the principles of 'Rethinking Construction'. Contractors and clients were encouraged to put forward examples of good construction practice known as Demonstration Projects focusing in particular on ways of improving Egan targets. This in turn led the development of the key performance indicators for the industry, regional networks through its Demonstration Projects Cluster Programme and regional Rethinking Construction events. The Construction Best Practice Programme sought to raise awareness of best practice, gain commitment to change, facilitate the sharing of knowledge and equip construction companies and clients with the knowledge and skills to implement change. Its services included a help desk which companies could contact, a website to disseminate good practice, best practice workshops and a company visit scheme. Its many publications included case studies, company and project profiles and guides, directors' briefings and information on the learning by doing workshops. It also established a team of forty best practice advisers and helped companies in the construction sector make better use of information technology through a spin-off IT Construction Best Practice programme.

The 1990s was also a decade of reform in other countries, but with a different focus. For example in Australia, the Gyles Royal Commission into the Building Industry in New South Wales investigated criminal activity in the industry and claimed to have found 'evidence of widespread lack of integrity and probity amongst the management of contractors and others involved in the Industry' (RCBCI 1992: xiv). As a result of the First Interim Report of Royal

Commissioner Gyles, the Building Industry Task Force (BITF) was set up on 30 September 1991 as a multidisciplinary task force of lawyers, industry experts, financial analysts and investigators (who were mostly police) to act as a law enforcement agency in the building industry. Nevertheless, the primary mechanisms used to achieve this were award restructuring, codes of practice and public standards and policies to establish industry-wide expectations of behaviour (McGeorge and Loosemore 2003). Much of this reform was led by the Construction Industry Development Association (CIDA). The late 1990s also saw increasing concern over a number of problems such as the relatively slow uptake of IT and unacceptably high accident rates. These were seen to be contributing significantly to both the financial and social costs of production. Like the UK, Australia's federal government in its *Building for Growth* report (DISR 1999) advocated the use of demonstration projects, particularly in promoting the advantages of integrated IT solutions with traditional industry practices. The proposed initiatives offered by the NSW State Government in the *Construct NSW* (DPWS 1998) document were similar in nature and included a Construction Policy Steering Committee to monitor IT initiatives. But despite governments encouraging and promoting the use of integrated IT solutions, the take-up was low because they were ultimately unable to control the purchasing and business decisions made in industry. In the area of Occupational Health Safety and Rehabilitation reform, the NSW Green Paper and *Construct NSW* report placed a heavy emphasis on the importance of key performance indicators (KPIs) as a tool to drive substantial improvements in performance. In 2001 WorkCover NSW (the statutory body concerned with the regulation of occupational health and safety (OHS) in the state of New South Wales) introduced a major new regulation entitled the Occupational Health and Safety Regulation 2001 which replaced all thirty-six existing regulations in this area. The new regulation not only considerably simplified the law in this area but also placed obligations on employers to adopt a systematic and consultative risk management based approach to managing OHS. The concept of sustainable development also emerged with the NSW Green Paper discussing the development of an environmental management system, which would provide a basis for accrediting and monitoring contractors at both organizational and project levels. Similarly, much of *Construct NSW* based its strategy on the principle of extensive monitoring and accreditation of contractors working on government projects. As with most of the other state government initiatives, its effectiveness was limited to the relatively few companies working on government projects and, while progressive, was by no means a comprehensive solution to the problem.

Construction management research

Construction management research during the 1990s was also beginning to mature. In particular, Hillebrandt and Cannon (1994), Male and Stocks (1991) and Betts and Ofori (1992) indicated that construction firms had learnt little from the recession of the 1970s and that during the 1980s and early 1990s they had demonstrated very little strategic thinking. Construction firms, it seemed, had become complacent and inward looking and had taken their eyes off emerging overseas markets. Strategic planning within the 1980s' construction industry had also been focused on reactive and short-term actions designed to improve internal efficiency (such as laying off staff). This had damaged the longer-term competitive position of firms and their capacity to grow into new markets. In terms of structure, the industry still had relatively high levels of private ownership and family control (although this was changing in the late 1990s as larger firms sought greater public ownership). There was also a high level of fragmentation (large numbers of SMEs working for a few large firms) and firms produced relatively heterogeneous products which were not particularly distinct (and thus were highly substitutable).

Using Porter's theories, researchers in construction started to reveal the main differentiators for contracting firms. In particular it became apparent for the first time that construction was a knowledge-based industry and that most competitive advantages were derived from knowledge-based factors (not from price). Unlike manufacturing, where technology-based innovations could provide significant competitive advantage, most construction firms were low-tech. Furthermore, the relatively limited technologies used in the construction industry were available to all firms. Therefore, it followed that the competitive edge for most construction firms resided in the competencies of their human resources and how technologies were used, rather than the technologies themselves. Furthermore, since most construction firms employed very few operational staff and subcontracted most of their work, much knowledge now resided in their supply chain of subcontractors (which were also available to competitors). So it was through the quality of 'relationships' with those subcontractors and through the 'preferential access to knowledge' that this afforded, that competitive advantage was gained.

2000s – a decade of consolidation and reflection

The 2000s were characterized by continued globalization and technological advances, the decline of the West's economic dominance and the emergence of Asia (particularly China and India) as world economic powers. The strategic management literature during this period conceived organizations as complex adaptive systems and was informed by the knowledge-based view of the firm (Grant 1996). This promoted the view that knowledge was a firm's most strategically significant resource and that intangible knowledge-based assets were replacing tangible physical assets as the primary source of competitive advantage. Building on this, the idea of stakeholder management also began to influence the strategic management literature. Advocates of stakeholder management argued that the relationships a firm had with its customers, suppliers or employees (which is based on trust and reputation) were a unique source of 'social capital' and competitive advantage. A firm was seen as a complex, dynamic and interdependent network of multidimensional relationships with a wide variety of stakeholders, the quality of which could affect or be affected by its activities. Performance and competiveness depended on how well firms managed and nurtured these relationships strategically and how they were perceived to manage them by stakeholders. Cummings (2005) argued that while one aspect of strategy must be focused on competitive advantage by reducing costs, increasing sales etc., it was also increasingly about recognizing the value in a firm's relationships, patterns, culture and history and the tacit knowledge and social capital embedded with this. Building on these ideas, the literature in this period also reflected the idea of strategy as a 'portfolio of relationships'. In other words, firms did not need to restrict themselves to gaining competitive advantage from the assets they owned. Instead they could leverage other firms' assets through contracting arrangements such as joint ventures, licences and alliances etc. to provide a value-adding business service they may have otherwise provided internally but which was not seen as core business. This 'outsourcing' model became very useful when a firm wished to expand into a market without expanding its breadth of internal core capabilities. It was also useful if a firm needed to adapt to changing markets rapidly without having to build up new internal capabilities.

The above ideas and the developments of new ICTs formed the foundations of a new era of 'virtual' organizations which blurred corporate boundaries. It also introduced notions of 'economies of expertise' to complement the idea of 'economies of scale' that had guided previous strategic thinking. So the new unit of analysis for strategy in the 2000s was not the firm or the market as it was in the past, but an amorphous network of firms bounded by fuzzy boundaries which was not as easy to define or track. As Venkatraman and Subramaniam (1986) pointed out, strategy had become a network-centric concept and at the end of the decade the questions being asked were around the forces that shape these networks and how a firm leveraged value from it and positioned itself optimally to achieve competitive advantage.

Construction industry development

In the construction industry, firms during the 2000s refocused on their core business. They also focused on financial management and profits rather than turnover and continued tightening organizational structures to reduce waste, facilitate specialization and to increase workforce flexibility. As Green (2011) points out, during the 2000s construction firms faced both unpredictable workloads and increasing competition and responded by basing their corporate strategies on leanness and agility, leading to the 'hollowed-out' firm of casual employees as the dominant form of organising. The development and emergence of major new international markets and construction capacity in developing countries also led to new mergers and acquisitions which created some truly global multinational corporations, laying the foundations for a global construction industry. Furthermore, skills shortages in many countries, as a consequence of decades of under-investment in training, led to new global mobility in migrant construction labour and its associated challenges.

Construction industry reform

In terms of construction industry reform, the early 2000s saw reports coming 'thick-and-fast'. For example, 2000 saw the launch in the UK of a major initiative called the Local Government Task Force (LGTF) to promote the principles of Rethinking Construction in local authorities. Local authorities, as one of the biggest construction clients in the country, were seen as vital links in achieving the improvement targets required by Rethinking Construction. By focusing on the whole-life costs of a project, rather than cheapest initial tender costs, local authorities were exhorted to meet their 'best value' obligations, and to deliver high quality services to the people they served. The LGTF published advice and guidance to local authority practitioners, designed to maximize their efficiency and effectiveness. The idea was that by avoiding waste, duplication and dispute, the community's interests should be served better. Working closely with the Movement for Innovation and the Housing Forum, the LGTF focused attention on demonstration projects.

In 2000 the UK's Construction Industry Board (CIB) also reviewed its role and responsibilities resulting in a change of remit to develop policies that would lead to improvements in the quality and skills of the construction work force, the efficiency and profitability of firms, the quality of their construction and in the value-for-money which clients receive. From 2001 onwards the Strategic Forum was established as a successor to the CIB with six key sector representatives to promote and to monitor industry progress on six key areas (described as its Construction Commitments). These areas remain: procurement and integration; commitment to people; client leadership; sustainability; design quality; and OHS.

The Confederation of Construction Clients was also formally launched in December 2000, comprising the entire membership of the previous Construction Clients Forum with the support of the Construction Round Table. The aim of the Confederation of Construction Clients was to: encourage clients to achieve value for money through best practice; secure major measurable and consistent improvement in performance across the industry; and promote policies aimed at achieving a safe, stable and skilled industry which was competitive and competent.

Many reports continued to be produced by numerous bodies including the National Audit Office's *Modernising Construction* (NAO 2001) which essentially repeated most of the suggestions of previous reports in how the United Kingdom's construction industry could be modernized with a focus on major clients. In 2001, the Strategic Forum for Construction, chaired by Sir John Egan, was also established and tasked to accelerate change and maintain the momentum of the original Egan task force. In 2002 the Strategic Forum for Construction published *Rethinking Construction – Accelerating Change* (SFC 2002) which was intended to encourage the UK construction industry to continue to aspire to the targets set in *Constructing the Team*. The ever-present theme of the need for client leadership was reinforced in delivering a culture of 'continuous improvement' driven through integrated teams, greater attention to 'people issues' (especially health and safety); higher design quality; better use of IT; more R&D; improving sustainability and whole life performance; and improved town planning systems. In 2001, a report entitled *Constructing the Future* (FCAPP 2001) was also published, mapping out future directions for the UK construction industry in the twenty-first century in the areas of: 'smart' buildings and infrastructure; health and safety; supply chain integration; investing in people; improving existing built facilities; exploiting global competitiveness; embracing sustainability; better forward planning and industry strategy; and increasing investment returns through better risk management.

The profusion of repetitive reports and overlapping reform bodies discussed above created an unwieldy and confusing array of recommendations which were largely rationalized in 2003 by the creation of Constructing Excellence. With the intention of creating a 'single voice' for improvement in the built environment sector, Constructing Excellence merged bodies such as Reading Construction Forum, Design Build Foundation, Construction Best Practice Programme, Movement for Innovation, Local Government Task Force, Rethinking Construction, Built Environment (Be), Constructing Excellence and Construction Clients Group. It continued to run the demonstration project and best practice work started by the Construction Best Practice Programme and Movement for Innovation but also produced its own reports such as *Be Valuable* (Saxon 2005) and *Never Waste a Good Crisis* (Wolstenholme 2009).

Wolstenholme's (2009) report reviewed progress since the 1998 Egan Report and made further recommendations to the industry. Wolstenholme (2009) found that progress in achieving the improvement targets set out in *Rethinking Construction* had been variable and disappointing particularly in improving cost, time and predictability performance, and to a lesser extent defects and safety, although productivity and profitability had improved. Despite a decade of intense analysis, debate and pressure for change in the construction industry, reform had only been 'skin-deep' and that the need to continue to challenge traditional ways of working was as strong as it was a decade before.

Wolstenholme also pointed to pessimism about the future outlook for change within the UK construction sector and claimed that this pessimism came down to a number of self-reinforcing 'blockers' that had continuously stifled change in the industry. These blockers were the result of a lack of 'joined-up thinking' in government, industry and professions about how the built environment contributed to economic, social and environmental sustainability. There were also far too many industry bodies with overlapping, blurred and competing constituencies and objectives meaning that key messages from the industry were not only diluted but were often contradictory. Furthermore, for much of the supply chain, there were few business or economic drivers to deliver meaningful change. Wolstenholme argued that an attitude of risk transfer prevailed and that too many firms were willing to accept stable returns rather than attempting risky innovation which could potentially produce extra ordinary returns. And despite demands for clients to drive change, the relatively small impact of construction costs and outcomes on most clients' core business meant that in many sectors reform in construction simply did not matter to them. Consequently, most client business models remained focused on short-term gain and did not reward firms which could deliver long-term sustainable solutions. Essentially, buildings were seen by many clients as a commodity purchase rather than a long-term investment, meaning that they focused on the up-front costs of construction, rather than the value created over the entire lifetime of an asset.

In discussing the important role of clients in driving innovation, Wolstenholme also argued that many clients struggle to articulate what value means to them, to provide a clear brief that defines their requirements and are unaware of the potential value that integrated supply chains can bring. Instead they rely on consultants specifying traditional solutions through outdated and self-promoting procurement methods which may not be suited to client requirements. Furthermore, few suppliers, other than those involved in PPPs, had any continued interest in the operation of the built asset and therefore no incentive to raise quality standards. Indeed, as the economy turned down in the latter years of the decade, many clients, who had previously championed partnering, had hypocritically returned to traditional procurement routes which enabled them to exploit market competition in their own favour. It was clear that in many cases, the concept of partnership was one that was only employed when it suited the interests of clients and that the culture of self-interest remained firmly imbedded within the industry.

Finally, Wolstenholme also argued that the industry lacked leadership talent which could promote a positive image and vision of the industry to government and the public. This meant that the industry failed to attract a high quality labour force and to recruit talent from untapped minority groups such as women and ethnic minorities. Furthermore, driven by the interests and accreditation demands of narrowly defined professional bodies, construction education perpetuated traditional professional and disciplinary boundaries resulting in constrained thinking about new integrated solutions.

So during the 2000s, improvement in the UK's construction industry appeared to have been largely restricted to the five hundred or so demonstration projects monitored by Constructing Excellence and its predecessors. Outside these demonstration projects (which were largely irrelevant to the many small-to-medium sized enterprises which dominated the construction sector), there had been little change in industry performance. The demonstration projects, although interesting, had been unsustainable and atypical (one can't use the A-team on every project) and had measured success in fundamentally flawed ways (projects were self-assessed against simplistic criteria). As Wolstenholme pointed out, it seemed that most firms were paying 'lip service' to the Egan agenda and cherry-picking the behaviours they wished to adopt, based on their own self-interest. For example, while many clients said they wanted a best value solution, they continued to employ on the basis of the lowest tender price. Furthermore, it was not possible to clearly establish how far the improvements were attributable to the favourable economic conditions of the previous decade, as opposed to innovation and improvements in efficiency. The overall impression was of a few shining examples of progress against a backdrop of fairly entrenched behaviour. In 2009, there was still only around a 50/50 chance of a project coming in on cost or on time.

In Australia parallel reforms were taking place within the construction industry being driven primarily by the 2002 Cole Royal Commission into the Building and Construction Industry and the federal government's fixation on IR reform. Established by the federal government to inquire into and report on matters in relation to corrupt employment practices, the Cole Royal Commission was vehemently criticized by construction unions as an ideologically motivated political agenda to undermine their power. Nevertheless, in 2003, as a result of the Royal Commission recommendations, the government introduced the Building and Construction Industry Improvement Act 2005. This led to the controversial creation of The Australian Building and Construction Commission (ABCC) in October 2005, a construction-specific policing body with special powers to enforce IR and anticorruption law in the construction sector. The ABCC was given building industry specific laws that provided higher penalties for building industry workers for breaches of industrial law. It could also force construction workers (under threat of criminal liability) to give evidence in industrial relations disputes. In the view of the government of the day, this was believed to be the main source of the building industry's problems. The ABCC was heavily criticized by the unions for singling out the construction industry as a special case and subjecting its employees to coercive and special and unfair laws not seen in other industries.

In 2007 a change in government brought about a change in approach to the construction industry. There was a new view that fixing the problems in the construction industry could not be reduced to a simple formula. It was accepted that the focus of ABCC investigations had been on union activity rather than on the broader problems facing the industry and that a wider array of reforms was needed to improve practices and culture to ensure a cooperative and harmonious workplace. More broadly the new federal government argued for a focus on cultural change in the workplace and the building of new mutually beneficial partnerships between management and workers. This underpinned a desire for greater collaborative and cooperative relationships in the building and construction industry to address issues like OH&S, skills development and productivity. So in 2009 the government introduced the Building and Construction Industry Improvement Amendment (Transition to Fair Work) Bill 2009 amending the title of the Building and Construction Industry Improvement Act 2005 to become the Fair Work (Building Industry) Act 2009. Although the ABCC continued to operate until 2012, this effectively abolished the Office of the ABBC Commissioner and created the Office of the Fair Work Building Industry Inspectorate. It also removed building industry specific laws and introduced a number of safeguards in relation to the use of the power to compulsorily obtain information or documents.

In the late 2000s increasing awareness of climate focused construction industry reform towards transition into a low carbon future. For example, the UK's Climate Change Act calls for the net UK carbon account in 2050 to be at least 80 per cent lower than the 1990 baseline. According to UK Government's Innovation and Growth Team (IGT), which was tasked to consider how the construction sector could meet this highly ambitious low carbon agenda, this poses the UK construction industry with the largest change management program since Victorian times (IGT 2010). However, many barriers to achieving these ambitious targets exist, including: a lack of drivers for clients to demand innovation; the fragmented structure of the industry; a lack of collaboration between firms in the supply chain; and a lack of skills and knowledge of low carbon solutions in the design, construction and operation of buildings. Green (2011) points out that these barriers are essentially the same as those repeatedly identified by reports over the previous thirty years, suggesting they are institutionalized within the industry. Therefore, Green argues that to meet this challenge, it logically follows that a very different approach to industry reform is needed in the future if the industry is to successfully transition itself to a low carbon future.

In Australia, reform towards a more sustainable low carbon construction sector was driven by the Federal Government's Built Environment Industry Innovation Council (BEIIC). Established in 2008 and ending in 2012, the BEIIC acted as an advisory body to the Minister and an innovation advocate for the industry. It's philosophy was that innovation is an outcome of an efficient industry, not an end in itself. Therefore, BEIIC could only ever be a catalyst for further action and the role of government was to provide leadership through clear targets and an appropriate regulatory environment to support the construction industry on that journey. Central to BEIIC's agenda were innovation challenges like climate change, sustainability and industry competitiveness, regulatory reform, workforce capability, skills needs and access to new technologies. Australia faces a challenging combination of a rapidly ageing built environment, future resource constraints, changing demographics, regional competition and climate change. Key innovation challenges for the future include addressing energy and water use efficiency and housing costs in light of the new Australian Carbon Pollution tax and trading scheme. While there have been some positive signs of increased innovation in the Australian built environment industry to address such challenges, the BEIIC argued that building a stronger culture of innovation would help to address them by developing new ways of working smarter rather than harder. Key initiatives to meet these challenges arising from BEIIC included: Innovative Procurement Policy and Processes; Green Skilling for the Retrofitting Industry; Government Urban Policy and Programs; Innovative Built Environment Demonstration Projects; Innovative Tools and Technologies; Improved Coordination Across Government for the Built Environment Industry; Principles for an Innovative Regulatory Framework; and Community Engagement Strategy to Promote Sustainable Practices.

In the area of procurement reform, the BEIIC considered that government leadership was crucial to setting the right framework for industry action. One of the easiest ways to show the right direction is for government to take the lead in procurement policy. To this end, key practical recommendations include expanding the principles of alliance contracting, collaboration and integrated supply chains to a broader range of projects. The BEIIC also argued that major clients should base their requirements on performance specifications rather than prescriptive descriptions of specific methods and materials. Performance parameters need to be specified in very broad, high-level terms so as not to dictate industry solutions and provide the 'space' for innovation to occur. In particular, governments and major clients should build internal technical competencies to promote innovation. They should also explore and develop the potential of new technologies such as BIM to achieve more collaborative working environments and encourage a culture shift away from lowest-price short-term decision-making towards value-based, whole-of-life decision-making. Furthermore, in their role as regulator, governments should reduce duplication and uncertainty in approval processes. And, in setting standards, there is also an urgent need for greater standardization and more rigorous and mandatory sustainability targets and metrics to measure performance.

Construction management research

The construction strategy literature during this time challenged the widely accepted idea that strategic thinking was not possible in construction because of the industry's uniqueness, sensitivity to economic cycles, long-term volatility and variable demand. Langford and Male (2001) and De Hann et al. (2002) argued that the key to strategic management in the construction sector was to develop unique competencies which were suited to specific business environments. Since most construction firms had access to the same physical resources at broadly similar costs, a firm's distinctive capabilities were derived mainly from its relationships which create its unique knowledge base, its capacity to innovate and its reputation. Reputation was argued to be especially important for construction firms that relied on repeat business. Relational capital (tied up in a firm's business relationships) also determined the quality of distribution channels for construction services and minimized the need for expensive marketing and promotional campaigns to remain at the forefront of client advisors' minds. However, since most construction firms have broadly the same networks of relationships, it was the quality of those relationships and the way they were combined and used to generate and share knowledge which was the main business differentiator. Building strong relationships with key suppliers and business partners also minimized 'switching costs' (the costs of moving between suppliers) and 'transaction costs' (the costs of dealing with a specific supplier).

The construction strategy literature at this time also argued that the introduction of new procurement routes such as PPP and alliances presented firms with new innovation opportunities in land assembly, financing, integration and managing the life-cycle of an asset. PPPs blurred the boundaries between private sector and public sector clients and fundamentally changed the relationships and risk distributions between key players in the construction industry, presenting new opportunities of benefiting from longer-term and more stable relationships. It also gave large contracting firms greater control over their competitive environment which was previously dictated by independent client advisors who chose the most appropriate procurement route (normally the one which gave them most power). With these changes, contracting firms moved upstream in the project life-cycle by offering advice and input to the preconstruction phases but also more downstream by offering more integrated finance, design, construction and operational services. This has meant that many construction firms became much more powerful, specialized and connected to the businesses of their clients, recruiting specialist staff from those sectors and developing specialist functional divisions to manage them.

Towards the end of the decade, Green et al.'s (2008) review of competitive strategy in the construction sector argued that traditional models used to understand the competitive strategy of construction firms had focused on exogenous (external) factors, whereas contemporary views of the firm emphasized the importance of endogenous (internal) factors. Using the perspective of dynamic capabilities (Teece et al. 1997), Green et al. (2008) found that the existing literature in construction management was very vague on what constituted key resources in construction firms and on what routines were needed for a firm to adapt to a changing environment. Green et al. argued that there had been a prevailing tendency to treat the management of key resources (in particular labour) as a static process rather than something that needed reconfiguring in response to changing business environments. According to Green et al. (2008), the construction reform agenda had largely neglected contemporary developments in strategic thinking and had uncritically adopted ideas from manufacturing which appeared to offer universal solutions to complex environments which instead required a more thoughtful approach to adapting organizational routines. Green et al.'s research also suggested that construction firm business strategies tended to develop organically and informally over time. Rather than being highly pre-planned as predicted in contemporary strategic management literature, construction firm strategies tend to emerge from the bottom-up, as a collective but contested endeavour of many people (often unsanctioned) from across different parts of a construction firm which may be in competition with each other for resources. While some strategies are initiated at board level and imposed top-down, the most successful business strategies evolved bottom-up, from opportunistic experiments and projects initiated by entrepreneurial employees. Green et al.'s research found little evidence that construction firms use the many formal strategic planning models or techniques which have been developed in the strategic planning literature. Furthermore, there was little evidence of corporate strategies being based on explicitly, formally and rationally planned processes, ploys to outmanoeuvre competitors or on a rational analysis of a firm's environment and matching resources to it.

Rather, Green et al. indicate a tendency for boards to retrospectively endorse strategies proven to be successful and to rely on the importance of relational and reputational assets with clients to generate new opportunities. Finally this research pointed to the strong constraining influence of path dependency in how firms explore new opportunities and on the close relationship between employees and their business environment to identify them.

The future

Many influential organizations attempt to predict the future on a regular basis in order to guide business on its response. For example, in its *Vision 2050: New Agenda for Business* report, the World Business Council for Sustainable Development argues that the future holds 'vast opportunities in a broad range of business segments as the global challenges of growth, urbanization, scarcity and environmental change become the key strategic drivers for business . . . ' (WBCSD 2012: 3). The World Business Council for Sustainable Development is a CEO-led organization of forward-thinking companies which seeks to galvanize the global business community to create a sustainable future for business, society and the environment. The report, which is a summary of conversations with CEOs and experts from over two hundred companies across twenty countries, predicts that new opportunities for firms will emerge around developing and maintaining low carbon cities and infrastructure but that these will also be accompanied by a new myriad of new risks which we have never seen before. The World Economic Forum (WEF) also produces an annual Global Risk Report which attempts to classify the risks posed by a range of constantly shifting social, technological, environmental, geopolitical, economic and technical trends. The World Economic Forum is an independent international organization committed to improving the state of the world by engaging business, political, academic and other leaders of society to shape global, regional and industry agendas. Its mission is to facilitate 'entrepreneurship in the global public interest' and in its latest report, the top risks include: global pandemics; religious fanaticism; chronic disease; extreme weather; antibiotic resistant bacteria; organized crime; illicit trade; unilateral resource nationalization; chronic labour market imbalances; infrastructure neglect; climate change; and the unintended consequences of new technologies (WEF 2013). While the World Economic Forum focuses on global trends, most countries also produce their own national and regional forecasts such as the CSIRO's analysis of global trends, shocks and scenarios which could specifically impact Australia (CSIRO 2010). Based on an analysis of over 100 trends by forty leading Australian scientists, the five interrelated trends predicted to affect Australian business in the next decade are: limited global resources; the increasing personalization of products and services; ageing and health problems in the developed world versus high fertility rates and food shortages in the developing world; urbanization and increased labour mobility; and digital and natural convergence as social interactions, information systems, transactions and sensory systems are increasingly replicated on the internet. While insightful and useful, the generic nature of these reports makes it difficult for businesses in specific sectors to construct a strategic response. So in full knowledge that we are about to attempt the impossible, the following sections reviews some of the main highlights of the 'futures' literature to speculate what the future may hold for the construction industry. We then discuss how contemporary strategic management theory may help construction organizations adapt and respond in an appropriate way.

Political and economic trends

In 2012 the annual meetings of the International Monetary Fund (IMF) and the World Bank provided insights into two profound changes taking place in global politics and economics. First, political and economic power has been moving inexorably east and second, we are in a prolonged period of economic instability.

As Sachs (2012) points out, while Western Europe and the United States have dominated political and economic agendas for the last century, we are now moving into an Asian century and a world with many regional powers and no obvious global leadership. For example, Hawksworth and Cookson (2008) project that over the next twenty years the emerging E7 countries (China, India, Brazil, Russia, Mexico, Indonesia and Turkey) will grow by 50 per cent more than the current G7 countries (US, Japan, German, UK, France, Italy and Canada). By 2050 the Brazilian economy will be larger than Japan and the Russian, Mexican and Indonesian economies taken together will be bigger than the collective economies of Germany, France and the UK. Other countries with high growth potential rates into the future include South Africa, Nigeria, Philippines, Bangladesh, Saudi Arabia, Malaysia, Thailand, Pakistan and Iran (assuming political problems do not pose a barrier to growth in the future).

The growth in these developing countries is reflected in the rapid increase in international jet travel which Boeing forecasts to grow from 4,621 billion kilometres in 2008 to 12,090 in 2028. Boeing also predicts that the sectors with the greatest annual growth rates will be Africa to South East Asia (9.2 per cent), within China (8.6 per cent), within South West Asia (8.7 per cent) and within South East Asia (8.1 per cent). These new travel patterns will reflect and further fuel the process of globalization and a shift of manufacturing capacity and capability from the developed to the developing world. In response there will also be a rapid growth in the service sectors of developed countries. For example, in 1970, services represented 55 per cent of the Australian economy. Today it represents over 75 per cent and there will be a ripple effect over coming decades as industry and society adjusts (CSIRO 2010). Finally, as Wade (2012: 1) warns, while the dawning of the Asian century presents many opportunities for Western business, it 'won't be one-way traffic'. For example recent research estimates that in Australia alone, over 80,000 service-sector jobs have been moved overseas in the past four years and it warns that hundreds of thousands more are predicted to go in the next three decades. However, Dr Ken Henry

AC, a prominent Australian economist, public servant and special advisor to the Australian Prime Minister, who authored the Australian government's 'Asian Century' White Paper, argues that people must get over their hostility to offshoring and outsourcing. Instead, companies in countries like Australia must see themselves as regional rather than nationalistic and see offshoring as a source of productivity growth rather than threat. It is clear that to capitalize on the economic shift of power we are seeing, companies will need to be adaptable, flexible, resilient and creative in their thinking.

The second profound change taking place in global politics and economics is the new era of economic instability which is upon us, owing as much to limits of growth as to financial turmoil in global markets. The double-dip recession that has so badly undermined European economies means that they are expected to be in contraction for the next decade at least. And to fund their regular bailouts, these countries are looking towards South East Asia, the Middle East and the BRICS countries (Brazil, Russia, India, China and South Africa) to provide the necessary resources. In return, these countries are demanding more power and votes on the IMF to which Europe will have to agree. In the first bailout, these countries secured a 6 per cent increase in voting rights. In subsequent bailouts they are demanding more, arguing that this reflects long-term trends in international trade which has seen Europe's share of global GDP fall from 31 per cent in 1980 to 20 per cent in 2011 to a predicted 17 per cent by 2017. The US is also in decline with its share of global GDP falling from 25 per cent to 19 per cent over the same period and being predicted to fall to 18 per cent by 2017. In contrast, the contribution of Asian developing countries to global GDP has risen from 8 per cent to 25 per cent over the same period and is predicted to grow to 31 per cent by 2017. Asia has become the new global centre of manufacturing challenging the West's self-image of economic dominance and forcing it to rethink its trading relationships outside its once comfortable and self-sustaining regional trading-blocks. Countries like Australia, which are located in the South East Asia region but which have had traditional ties with the West, will be profoundly affected by these changes of economic power.

The uncertain question of how South East Asia will evolve will present firms and countries with difficult choices around allegiances. For example, no one currently knows whether South East Asia will emerge as a politically integrated region or a disparate collection of emerging countries. The likelihood is, that having seen the crisis that has engulfed Europe's attempt at regional integration and common markets, South East Asian countries will be far more cautious about integration than they may have been five years ago. Furthermore, at the same time as growing in economic importance, these emerging countries seem reluctant to assume their share of global leadership, potentially leading to a world of five or six major powers. And then of course there is the question of whether China and India can continue their race to catch up with the developed world. Most commentators expect the rise of China and India to change the landscape of the global economy for ever and the OECD (2010) predicts that China will replace the US as the world's leading producer of manufactured goods in the next five to seven years and will be around 130 per cent bigger by 2050. Similarly, India is predicted to continue to grow its expanding industrial and service sectors and is anticipated to be around 90 per cent of the US economy by 2050. However, in his book *The Next Convergence* Nobel Laureate economist Michael Spence warns of the challenges facing India and China in making this transition. In recent years, both countries have seen a stalling of structural transformation, where labour intensive export sectors which have driven growth are seeing wage growth and reduced competitiveness. And both countries have yet to fully dismantle the pre-reform cultural legacies of communism and the raj and cultural barriers to growth which prevent the same levels of education for women as for men. Furthermore, as Hawksworth and Cookson (2008) predict, China's growth could be threatened by a significantly ageing population due to the one child policy (although India's working age population is expected to grow at a healthy rate). The past experiences of Japan and other Asian Tigers suggest that China's high levels of savings investment-driven growth will eventually see diminishing returns as income levels approach OECD levels and as its ageing population cashes in its savings to fund retirement.

The global financial crisis has highlighted like never before the interconnected world in which we live and the need for clear leadership and corporation in dealing with the challenges of this generational shift in economic power. It has also demonstrated that we have a poor understanding of the interdependencies that exist between critical global sub-systems and of the 'cascading uncertainties' which they can produce. A relatively small problem in one part of the global economy can grow in proportions and have major impacts on another. So it is difficult to know where the next major shock to the global economy will originate. For example, one area of emerging concern is the rapid rise in global commodity prices due to growing global demands for growth from developing nations in an environment of finite resource supplies. There is also a growing sustainability and environmental crisis which is leading to increases in food prices and reduced water availability due to excessive unmanaged use of fresh water from rivers and aquifers. It is clear that if disaster is to be avoided, the world needs to pull together even if there is no obvious global leadership in the future.

Another consequence of the recent economic turmoil and the excesses of corporate governance that precipitated it is a surge in regulation and legislation governing business activities. These are requiring businesses to better disclose their activities and to consider the interests of a wider range of stakeholders in their business decisions. A recent survey by Deloitte found that 62 per cent of firms were not confident they could comply with this new surge of regulation (Risk 2009). Growing regulation is also leading to greater risk aversion and concerns in business circles about growing personal liability, scrutiny of directors and a greater willingness by courts to award damaged for malpractice and non-compliance. And as globalization gathers pace, there are also concerns about liabilities associated with trade relationships with overseas countries which do not have the same regulatory controls that developed countries have. A recent Australian survey estimated that only 12 per cent of firms comply with Australia's anti-corruption legislation and the Berlin-based anti-corruption watchdog Transparency International's 2005 Global Corruption report labelled construction as consistently the most corrupt industry in the world (Transparency International 2005). In 2008 the construction industry topped its Bribe Payers Index and according to the report, construction companies are the most corruption prone when dealing with the public sector and are the most likely to try and influence the policies of governments. Large and complex construction and engineering projects are fertile ground for hiding bribery and corrupt business practices.

So governance is an emerging issue which merits particular attention into the future, especially for global firms which operate across international borders. Social and political developments in society are creating unprecedented levels of scrutiny and awareness of company activities and greater expectations that firms disclose their economic, social and environmental performance. The public also increasingly expects that firms will consult stakeholders in the process of doing so. And if organizations do not comply with these expectations then the media and public are increasingly empowered and willing to expose them and do something about it. As Loosemore and Phua (2011) predict, given the size, scope and impact of construction activity on the communities and the environment the concept of corporate social responsibility is likely to be a major driver of construction sector strategy into the future. The strength of this trend should not be underestimated and there is evidence that there are significant reputational benefits to be gained from being perceived as an open, trustworthy and socially and environmentally conscious organization. For example, Oury (2007) found that a recent study found that 70 per cent of final year undergraduate students said that a company's ethical record was crucial when choosing an employer. Kotler and Lee (2005) found that 55 per cent of MBA students would accept lower wages to work for a company that believed in something (these tended to be the higher achieving students), that 76 per cent of these MBA students would be more likely to stay in a job that was related to a cause, that 78 per cent are more likely to buy a product associated with a cause and that 80 per cent of customers would be likely to buy from a company with a positive image.

In response to the above, broader and better performance metrics are likely to be an important dimension of business change in the future. The demand drivers for increased and broader-based performance reporting include: the investment community, where there is a growing expectation of corporate social responsibility and sustainable practice; legislative and regulatory requirements which seek to provide clarity and certainty to the public and government around licences to operate; recognition by business themselves that sustainable practice is good and profitable; and increased expectations of the general public that businesses should be more accountable and transparent.

Technological trends

In a recent interview about the future of technology, IBM's new CEO, Inni Rometty, said that history has produced only three computing eras. The first era during the 1950s produced machines that could count, tabulate and calculate using the punch cards machines which Thomas Watson first produced. The second era of computers, which began in the 1960s, were programmable and this era continues to this day with every computing device including the iPad. The next 'cognitive era' will produce machines that understand language and generate hypotheses to learn, refine and perfect their responses over time. IBM believes that the predictive capabilities of this new technology could transform business, especially those which are struggling to manage large amounts of data. It also opens the way for a new generation of robots which will act like 'virtual humans' serving as customer support slaves which can engage in real conversations with customers, through analysing conversation patterns and the context and profile of the person they are interacting with. For example, an Australian company called MyCyber-Twin offers companies online workers that can be created and trained to perfect tasks in a couple of hours. As the founder of the company Liesl Capper argues, 'the idea is that anyone should be able to have a virtual staff member, a human-like slave, on their website, whether you are a plumber, a small bank, or whatever' (Steffan 2012: 7). She goes on to explain that unlike real people, virtual staff members can work 24/7 without any absenteeism, without any complaints and without requiring a single pay cheque. Furthermore, experience has shown that people are surprisingly willing to treat robots like real people and form relationships with them. Capper predicts that in a decade 15–20 per cent of human interaction could be with people-robots, especially since Apple mainstreamed the idea by launching its iPhone assistant called Sirim opening the business community's eyes to the huge value of the technology.

Despite continuous advances in technology, a report by the Construction Industry Institute into the use of technology in the construction sector concluded that construction has always lagged other industries in this area (CII 2008). There are many reasons for this such as the bespoke one-off nature of buildings, the unpredictable production environment in which they are produced, traditions and restrictive practices which are threatened by the introduction of new technologies, a lack of investment in R&D and the labour-intensive nature of the construction process. CII (2008) also found that existing skill levels in the industry are not advancing fast enough to use this new technology, that there are problems of synergies and interoperability with existing technologies and that work processes are very slow to change to accommodate the new technologies. Nevertheless, evidence shows that even the low rate of technology uptake has increased productivity in the construction sector considerably over the last decade (estimated between 30–40 per cent). In particular, construction automation, integrated project information systems and automated material identification and tracking systems have had a major impact and there are a range of new technologies on the horizon which are promising even greater advances into the future. Hooper and Harris's (2010) predictions of technological advances which are already starting to shape the future of the construction industry and built environment include:

- Business information modeling and virtual reality which enable project teams to collaboratively build, experience and test a virtual building, precinct or even city prototype before work starts.
- Wearable and ultra-mobile computers and APPs which utilize touch screen technologies.
- GPS guided plant and machinery.
- Extra high strength lightweight concretes which use basalt fibre or glass composite reinforcement.
- Light transmitting concrete containing optical fibres as reinforcement.
- Augmented real estate, which enables stakeholders to see how new buildings fit into and interact with their surrounding built environment.
- Carbon negative cements which absorb CO_2 during their production and life.
- Smart windows which produce electricity and which can change the opacity of glass to filter light.
- Rotating buildings which can maximize solar energy gains.
- Self-healing concretes which incorporate un-hydrated cement which react with water and carbon dioxide.
- Buildings that breath using automated ventilators and prevent the need for air conditioning.
- Sensors imbedded into buildings and infrastructure that can relay performance data back to central computers which can automatically adjust design settings.
- Stereolithography which allows building components and even whole buildings to be manufactured through new concrete and polymer printing technologies on site.

Brandon (2010: 29) predicts that we are moving towards a 'machine-dominated' construction industry. And if Moore's Law holds true, in five years computers will be ten times faster or ten times cheaper than they are today. In particular, wireless and gaming technologies appear to be emerging as a real game changer, fundamentally altering the relationship between businesses and their customers. The world's online gaming market has more than tripled over the last five years and innovative companies in the construction sector are already experimenting with this technology as a new way to construct virtual interactive dynamic models of buildings to communicate better with their clients and customers. Some construction companies are also developing site-based simulation games to allow their project managers to play with different logistical configurations to optimize project management strategies. And as more people use social media, the lines between our professional and personal lives will become increasingly blurred. Three hundred and fifty million people worldwide now use Facebook exchanging over 30 billion pieces of content every month and new ways of online networking will evolve to offer businesses enormous potential to network, communicate, consult and collaborate in real time with internal and external stakeholders. New online tools will also enable business to engage with customers in completely new ways never imagined before. For example, on 10 December 2011 the world's biggest online retailer Amazon asked its millions of online customers to use its new smartphone barcode scanning application to scan the prices of any three competitor products and get a discount off Amazon products for doing so. As Ferguson (2012) points out, the aim was to provide Amazon with information about its competitors' prices but also to alert customers to the fact that they can use new technology to compare prices of their competitors' products.

Technological advances will also enable firms to customize and personalize their products and services like never before. As CSIRO (2010) points out, those firms that are able to personalize their services and products are likely to gain more loyal customers. And while retailing is arguably undergoing the most dramatic technological transformation in customer habits, construction is also being affected by these trends. For example, with the growing use of business information modeling firms should be able to interact with customers and business partners in totally new ways through integrated design and virtual optimization of building designs in a collaborative environment. GPS and spatial information systems are also likely to revolutionize many parts of the construction industry – particularly the work of design and planning professionals. For example, GPS systems are being used by firms to set out sites more accurately and to help automate excavators to cut more accurately with less waste. It might not be long before some machines will drive themselves.

As well as allowing business to engage with clients more effectively, new ICTs will also allow business to engage with the public more effectively. For example, Jacob Colker is co-founder and CEO of Sparked.com, which has leveraged the latest technological advancements to revolutionize the way business and community organizations collaborate to help the world's disadvantaged. Sparked.com allows busy professionals to use their skills in creative design, marketing, strategy, media relations, technology and more, to help non-profits and governments around the world solve global challenges such as natural disasters. This has created a new virtual generation of 'micro-volunteers'.

As the use of ICTs grows, we are also facing an explosion of digital information unrivalled since the advent of the internet, leading to increased information overload for managers and unprecedented demand for information management services. A recent survey of 124 managers in Australia, Hong Kong, USA and UK showed that 72 per cent of respondents were suffering significant time losses in searching and processing information and about 37 per cent experienced information overload on a daily basis (CSIRO 2010). Side effects included stress, reduced decision quality, frustration and tiredness which reduced productivity. This trend is unlikely to stop since there is currently a 40 per cent growth in global information generated every year and global spending on cloud computing is predicted to grow by 300 per cent over the next year, because it offers a much cheaper way for businesses to store information and keep up to date. This will create big challenges for companies in managing cyber-crime and information security and communications. And history suggests that many firms in the construction industry will struggle to understand how to share and secure the large amounts of new data that will be generated in the future about how buildings are built, function and perform. One can imagine what would happen if terrorists got hold of information about water supply and ventilation systems, when thousands of people can inhabit a single building.

The increased use of smart phone technology also presents new risks and opportunities for business in the future. For example, on the upside some firms are experimenting with smart phone technologies to enable their customers to interact with buildings more remotely and effectively. Other construction firms are adapting their employees' mobile phones to allow them to automatically collect safety risk data as they walk around a site and to track the location of workers on large remote projects. However, there is also a downside with this technology. In particular, there are ethical issues around tracking employees and since large amounts of data can now be downloaded onto smart phones it can also be lost or stolen far more easily than in the past. The sheer variability of platforms and types of phone used can make the task of managing these risks extremely difficult.

Other emerging technologies which might transform the construction industry include new flexible optical fibres which can be incorporated into materials like concrete to provide them with optical light emitting properties. And using stereo-lithography it is now possible to 'print' custom building components or even whole buildings in situ in concrete or polymers using an onsite computer linked to a design model and a building material cutter or printer. So far, the construction sector has been resistant to increased automation and off-site fabrication, despite many proven benefits such as improved safety, increased quality and cost and time efficiency. For example, in Sydney, Australia, Brookfield Multiplex's One Shelley Street development used off-site manufacturing to avoid 50,000 hours of work that would have been done at a height. This left only 3,500 hours of on-site rigging leading to a 75 per cent reduction in safety risk (the fatality risk in a factory setting is about half that on a construction site). In Australia, the Hickory Group has developed an innovative structural building technology called the UB® System, which accelerates on-site and off-site construction programs by up to 50 per cent, minimizing material and energy waste, whilst maximizing quality and safety. In China, Broad Sustainable Building is planning to build a 838m high skyscraper called Sky City in just three months using innovative prefabricated technologies in its own factories (compared to five years to build the 828m Burj Khalifa in Dubai). Using this approach, Broad Sustainable Building plans to have 95 per cent of the new skyscraper manufactured off site before the ground is broken. Their patented designs and technologies produce buildings which consume 20 per cent of the energy of conventional buildings and cost less than half the money. For example, the Sky City development is estimated to cost US$628 million compared to US$1.5billion for the Burj Khalifa in Dubai.

New energy technologies are also set to revolutionize the way we live. The world's growing population, rising per capita energy use and ageing coal-fired electricity generation capacity means a huge amount of investment will be required in alternative energy generation. Huge capacity additions will be needed to deliver a new energy mix which will include: coal, gas, nuclear, hydro, biomass, wind, solar and geothermal. For example, by 2050 it is estimated that micro generation could provide up to 40 per cent of all UK energy needs. And a study at Stanford University by Jacobson and Delucchi (2009) found that to provide base-load electricity from sustainable sources, the world would need a geographically interconnected web of 3.8 million 5MW wind turbines, 49,000 large-scale 300MW concentrating solar power stations, 40,000 large-scale 300MW solar photovoltaic power stations, 1.7 billion domestic scale 3 KW rooftop photovoltaic systems, nine hundred 100 MW hydro power stations, 5350 geothermal 100MW plants plus many thousands of small tidal and wave installations. Less than 1 per cent of this infrastructure is currently in place and the investment would need to be over $100 trillion over twenty years (excluding transmission) to make up the difference. Alternatively, if one opted for nuclear power then 17,000 nuclear plants would be needed to meet worldwide demand. So while a rapid change to renewable is unlikely, it may sometime become a necessity and there is no doubt that there will be a significant switch from fossil fuels to renewable over the next decade. The only question is by how much and how fast.

Many of the technological advances we are likely to see in the future will be driven by the demand for more sustainability business solutions. For example, it is likely that taller buildings will be needed to increase urban density, reduce car dependency and reduce energy consumption. London only had two skyscrapers in 1999 but by 2014 it is anticipated there will be 24. And these buildings will need to be energy generators rather than energy sinks. To this end, researchers in Australia are experimenting with ink jet technology to spray solar cells on to curtain walling. Already smart metering and new monitoring and valve technologies can provide real time feedback to users to help them change their energy and water consumption habits. Bio mimicry is also enabling designers to replicate natural systems and reduce our consumption of natural resources. And by using nanotechnology, engineers are experimenting with smart materials such as dust that can be mixed into concrete to monitor building performance and advanced paints that can clean themselves, absorb air pollutants in offices or even record the conversations of building inhabitants. Also, given the extremely high surface area to volume ratio of nano particles (a nano metre is one billionth of a metre), it is now possible to improve insulation products in a fraction of the volume occupied by current insulators. Vast increases in strength can also be achieved using this technology. For example, using nanotechnology Nobel Prize winning scientists Professors Andre Geim and Konstantin Novoselov have recently developed a new super strong material called graphene which could be as transformative to materials science as the development of polymers was in the twentieth century. Graphene is stronger than diamond, hundreds of times more conductive than copper and is so flexible that just one gram could stretch to cover several football pitches. It is already being used by IBM and Samsung as a possible replacement for silicon in computers but as Hannaford (2010) points out, graphene could also prove to be the ideal building material and lead to the development of super strong super lightweight structures. In other applications of nanotechnology, new piezo-electric fabrics are being developed which generate energy from movement and from light. The first applications will be as coatings on laptops, gadgets, cases, bags, tents and cars but there are also obvious applications to the skins of buildings which could generate energy from solar and wind movement. Although there are claimed health risks associated with nanotechnologies, the potential market in construction is immense and is predicted to be US$1.75bn by 2025 (Rushton 2010).

Finally, transport technology will also dramatically change the way we travel and in turn the urban environment in which we live. For example, in Tokyo, one of the world's largest metropolises of over 30 million people, commuters can travel over 200km to work in less than forty-five minutes. In contrast, for commuters in Sydney (population approximately 5 million) the same journey could take over three hours. The implications of these new transport technologies for cities are immense because they make possible the idea of efficient megacities. A megacity is broadly defined as a city with a population in excess of 10 million people. In 1800, only 3 per cent of the world's population lived in cities, a figure that rose to 47 per cent by the end of the twentieth century. In 1950, there were 83 cities with populations exceeding one million; by 2007, this number had risen to 468. The UN forecasts that today's urban population of 3.2 billion will rise to nearly 5 billion by 2030, when 60 per cent of people will live in cities. By 2025 Asia alone is predicted to have at least 10 megacities, including Mumbia (33 million), Shanghai (27 million), Karachi (26.5 million), Dhaka (26 million) and Jakarta (24.9 million people).

Social and ecological trends

The World Business Council for Sustainable Development (WBCSD 2012) predicts that the world's population will increase to over 10 billion by 2050 from approximately 6 billion in 2000, the vast majority of this growth occurring in developing countries. With an associated growth in global middle classes, this will lead to an acceleration of resource consumption per capita and higher expectations around health, education, income, material possessions, democracy, governance, security and environmental protection. The global governance and policy responses to these shifting demographics have typically happened in silos and have been limited by short-term, localized political pressures and self-interest. Consequently, they are likely to fall short of the commitments needed to alleviate these pressures. And as internal conflict and climate change reduce the capacity of Third World nations to carry their expanding populations, the global migration of labour (legal and illegal) to developed countries is likely to grow. For example, it is anticipated that at current rates of migration, Australia's population will grow from about 22 million in 2009 to 36 million in 2050 (SMH 2010). Today, 16 per cent of the Australian labour force is from a non-English speaking background and the Australian construction industry is one of Australia's largest and most culturally diverse, directly employing about 9.2 per cent of the working population (ABS 2009). While the benefits of migration include a continually growing economy and diverse workforce, the downside includes poor integration into societies and workplaces and increased stress on existing infrastructure and cities and scare resources such as energy and water. Cross-cultural researchers have found that capturing the potential productivity benefits of cultural diversity is complex and challenging (Dunn and McDonald 2001; Pedersen et al. 2005). It is likely therefore that many counties and industries like construction will struggle to integrate different cultural groups into a harmonious and productive workplace.

Increased urbanization and population growth, particularly in developing countries like China, India, South America, the Middle East and South East Asia, will also place greater stress on our energy resources and infrastructure. For example, based on current projections, China will need to build an additional two cities the size of Shenzen (7 million people) every year until 2030 to meet its urbanization demands. Seventy per cent of this growth is expected to occur outside planning controls and on the margins of cities' fringes, placing extra stress on urban infrastructure. While presenting many new opportunities, badly planned cities can also present an ideal climate for social fragmentation, crime and disease spread, raising all sorts of important planning challenges. To keep these new and increasing congested mega cities working efficiently will require new efficient road and rail transportation technologies and systems to enable people to move easily and to reduce car dependence. This will be crucial. As Manning (2011) warns, peak oil is closer than we think. The International Energy Agency has warned that the world's crude oil production peaked in 2006 and that the supply of oil will fall behind demand for the first time by 2014 and that oil will soon hit US$200 a barrel, although the global financial crisis may postpone this inevitability for a short time. This warning stands against a backdrop where currently over 25 per cent of the world's population have no access to electricity and where it has been estimated that world energy demand could triple over the next decade as developing countries expand their economies.

If not addressed the social implications of fuel poverty will be immense. For example, by 2020 the UK could be importing as much as 90 per cent of its gas needs compared to 10 per cent today. So governments around the world need to urgently start looking for alternative sources of energy and the built environment has a key role to play in addressing these challenges. Not only can new and existing buildings be made far more energy efficient but they can be used to generate renewable energies such as solar and wind to compensate for reductions in traditional fossil fuel supplies in the future. For example, in Barcelona, available solar energy from micro generation alone is about ten times the city's energy consumption. And countries like Portugal are striving to generate 45 per cent of its total energy needs from renewables by 2010.

Climate change will also likely be a major driver of innovation in the future. The majority of scientists propose capping atmospheric CO_2 concentrations to below 450 parts per million with the desire to keep the global temperature increase to less than 2°C. This in turn requires global carbon emissions to be cut back to 60–75 per cent of 1990 levels by 2020. To this end, many countries are adopting mandatory or voluntary measures to control greenhouse gas emissions. For example, the UK government has set a target to reduce building emissions by 80 per cent by 2050 and Melbourne in Australia has set out plans to be carbon neutral by 2020. A large proportion of these emissions have been identified as arising from the built environment (some estimates place it at over 30 per cent of all emissions). Vast quantities of carbon emissions are generated by the construction production process, by the materials used during construction and in the electricity and transport required to enable buildings to function individually and collectively in the urban context. For example, it is estimated that on an annual basis, buildings in the United States consume 39 per cent of America's energy and 68 per cent of its electricity. Furthermore, buildings emit 38 per cent of the carbon dioxide (the primary greenhouse gas associated with climate change), 49 per cent of the sulphur dioxide and 25 per cent of the nitrogen oxides found in the air. So firms operating in the construction sector have a major role to play in achieving these carbon targets and will be forced to think about how they can do so by governments, shareholders, public activists and employees who want to see real efforts to reduce emissions. The construction sector will also be fundamentally affected as major users of high emission resources such as steel, aggregates, timber, cement etc. The prices of these materials will inevitably increase significantly in the future. With estimates of up to 80 per cent of a firm's carbon footprint imbedded in its supply chain, large companies are already putting pressure on their suppliers to reduce carbon footprints and emissions. Furthermore as WBCSD (2012: 26) points out, 'buildings are the most economical path to saving energy and reducing CO_2 emissions'. Energy issues have become a high priority for owners and tenants and new energy codes are being developed and enforced, new incentives for low energy are being created and energy measurement and labelling of buildings are increasing transparency. In the future, new buildings will need to be at least zero net energy (and ideally energy sources rather than sinks) and existing ones retrofitted to the same standard through integrated design, new technologies, new materials and new workforce skills. But while technology is important in meeting this challenge, an engineering solution is only a partial one. An important and often neglected part of the challenge will be educating employees, clients, customers and users of built environment products and services about living with climate change and adapting their energy consumption expectations and behaviours to suit.

Another social challenge for the future is population ageing, particularly in the OECD. For example, the proportion of the Australian population aged 65 years and over has risen steadily over the past two decades and is projected to rise further over the next 50 years (White 2012: 19). In 2012, the average life expectancy for Australian men was 81 years and for Australian women it was 85 years. In 1901 life expectancy was 55.2 years and there were 151,000 people aged 65 years and over living in Australia (4 per cent of the total Australian population). The proportion of the population aged 65 years and over is projected to rise from around 12 per cent today to 25 per cent by the year 2051 as the peak of the baby-boom generation reaches retirement age. This is not just an Australian problem but a global challenge to which business must respond. Projections by the UN show that approximately 35 per cent of the developed world's population will be over 60 by 2050 compared to 12 per cent in 1950.

One of the side effects of an ageing population is a falling workforce and some countries will see their working populations falling faster than others. For example, Japan expects its workforce to shrink by 60 per cent over the next twenty-five years and the dependency ratio of workers to retirees is projected to fall from 3:1 today to about 1.5:1 in 2030. China's population will start to fall in about twenty years and South Korea's in about ten years. The impact of this intergenerational shift into retirement is far from clear, and the changing lifestyle choices and behaviours of ageing households over the next twenty years are also largely unknown. But we do know that these profound changes will create a smaller and older workforce with changing patterns of work and lifestyles which will create new demands for flexible working. And we also know that many firms are totally unprepared for this coming shift in workplace demographics and its inevitable impact on relative productivity. Combined with under-investment in training over recent decades, these changing demographics will sustain severe skills shortages into the foreseeable future in many developed countries. Migration from developing countries which have high fertility rates will help to alleviate these problems but will bring social challenges of its own. And firms will need to look to new non-traditional sources of labour to fill the void such as migrants, women and the disabled. For example between 1975 and 2000 the female participation rate has increased by 14 per cent in New Zealand, 13 per cent in Netherlands and 12 per cent in Norway. However, construction remains a male dominated industry with significant barriers to career progression for women. The lack of gender diversity in the Australian construction sector is a persistent problem and after two decades of industry and government policy intervention to promote gender diversity, women have fallen from 17 per cent of the Australian construction workforce in 2006 (ABS 2006) to just 11.6 per cent in 2012 (EOWWA 2012). In 2012 the Australian construction industry was Australia's most male dominated sector. Women in construction not only fare poorly among technicians, trades, labourers and machinery operators (3 per cent), but also in professional and management roles (14 per cent). Men in the construction industry tend to dominate senior 'technical' 'fee-earning' careers while women congregate in junior, marginal and less well-paid desk-based non-fee-earning professions such as human resources and marketing. This segregation also contributes to the problem of female recruitment and retention. In Australia, women are leaving the construction professions almost 39 per cent faster than their male colleagues and a 2010 industry survey of 1,100 women construction professionals found 48 per cent were disappointed by the pace of their career progression relative to their male peers and that 33 per cent expected to leave the profession within five years.

Contemporary strategic thinking

How is the strategic management literature positioning itself for these trends? Strongly influenced by fallout from the global financial crisis, contemporary writers like Clegg et al. (2011) are questioning traditional approaches to business strategy which have focused on short-term shareholder returns. They argue that strategy is contextual, social and political and that recent events and challenges have highlighted, like never before, the intimate relationship between business, politics, ecology and society. In other words, strategists cannot be absolved from the wider social and the ecological implications of their decisions. Furthermore, in contrast to traditional thinking which sees strategy arising from the individual inspirations of great leaders, contemporary thinkers see strategy as something which is 'constituted and enacted organizationally'. In contrast to the past, contemporary strategic thinking has also come to recognize that due to power and politics in organizations, strategy does not always develop or get implemented as planned. Rather, it is influenced by the interpersonal struggles, individual ambitions, rivalries, egos and power relationships of those affected. So contemporary ideas about business strategy question Tayloristic top-down thinking but emphasize the 'emergent' and unplanned nature of the process which arises from a contested and constantly evolving interaction between an organization's plans and its employee's interpretations of them. The business strategist must also recognize that the future is unpredictable and that organizational agility is the key to success in adapting to the future as it unfolds. This means that rather than defining strategic plans that try and predict the future and set goals in great detail (as much of the traditional strategic planning literature would suggest), strategists should set broad goals and visions and encourage employees to use their initiative in achieving them. The underlying assumption of contemporary thinking in corporate strategy is that today's valuable resources may not valuable in the future and that it is impossible to know in advance which ideas will fail and which will succeed. Finally, a business has to consider much more than simply the competiveness of its products but also its relationships and communications with its customers. Insights about customer needs through focused customer segmentation (understanding different customer needs), targeting (through detailed analysis of segment profitability) and positioning (by implementing a customer-focused strategy), will be key to business success in the future.

To what extent have the above ideas on corporate strategy filtered into the research and discourse on strategy in the construction industry? Certainly Green's (2011) analysis of the construction improvement agenda points to the continued political influences of the post 1980s enterprise culture, first promoted by Margaret Thatcher. This, he argues, underlies the rhetoric of the construction improvement debate which on the one hand promotes a more socially responsible approach to management but on the other hand continues to promote 'best practice' management recipes which have their origins in Tayloristic thinking. As Green (2011: 353) points out, the fashionable 'sound bites' of collaborative working, integration, agility and leanness continue to drive the construction industry reform agenda in many countries. These vague themes persist because they have never been fully defined and can be used to legitimize structural changes in the industry over the last thirty years. The rhetoric of collaboration he argues cannot be understood in isolation from long-term trends towards outsourcing, subcontracting and workforce casualization which have created the 'hollowed out' firm. And similarly, the mantra of integration should be seen as 'a rhetorical corrective' to the problems stemming from the construction industry's fragmentation over thirty years. It is much easier, he argues, to make repeated calls for integration than to do something about addressing the modernist management philosophies of organizational flexibility and efficiency which continue to perpetuate it.

Reflecting these ideas, Dainty and Loosemore (2012) brought together a number of leading authors who have, through their own research and practice, begun to question strategic human resource management orthodoxies within the sector. Together, the contributions untangle the intended and unintended consequences of industry management practices for both those who work in the industry and the organizations which employ them. What binds these authors together is the recognition that although many industry reforms in the UK have alluded to the importance of people, they in fact treat people like any other resource to be 'efficiently' managed, or worse to be exploited like any form of capital in the cause of improved business performance. In a reflection of Clegg et al.'s (2011) criticisms of mainstream strategy research, Dainty and Loosemore (2012) argue that the field of construction management research has been widely rooted in so-called 'best practice' panaceas, underpinned by Porterian thinking and a reluctance to adopt critical perspectives which consider the socio-political and moral implications of industry strategic practices. It is argued that these established management practices are at the root of the seemingly intractable problems which confront the industry.

Conclusion

The aim of this chapter was to provide an overview of how firms in the construction industry have adapted over time to the changing business environment. We have discussed the main periods of change experienced by the construction sector in developed countries since the Second World War and reviewed the strategic management thinking that may have influenced organizational responses. Predicting the future is notoriously difficult and we make no attempt to do so here. But what is clearly evident from this chapter is that the future holds many new challenges which we have not faced before and that the construction industry must innovate if it is to adapt effectively. However, it is also clear from this chapter that over the last fifty years, changes within the industry have created a highly rigid structure and culture which is highly resistant to change. The limited success in reforming the industry over the last fifty years has demonstrated this. A new approach is clearly needed which addresses the drivers, enablers, impediments and impacts of innovation which we have discussed in previous chapters. And this new approach needs to enable the industry to better capture and harness the many opportunities which are presented by a more uncertain, dynamic and interconnected world. These are addressed in the following chapters where we explore the value of research from outside construction to develop some innovative solutions to these challenges.

5 Strategy and innovation

> The enterprise that does not innovate inevitably ages and declines. And in a period of rapid change such as the present . . . the decline will be fast.
>
> Peter F. Drucker

Introduction

Strategy is an entrepreneurial activity which involves thinking about the future. Traditional notions of strategy (often referred to as the 'rational' or 'positioning' view) conceive it as a planned, top-down and mechanical process. This typically starts with a mission statement which is the responsibility of the Board of Directors and should communicate the fundamental purpose of an organization and its reason for being. The mission statement should act as a rallying call to employees about the company's core grass-roots values, what it considers important, its key stakeholders and what distinguishes it from its competitors. Next comes the vision statement which should complement this by articulating the ideal position that the organization wishes to achieve and the means by which it wishes to do so. McKeown (2012) argues that a lack of vision is the single most significant impediment to innovation in business. First, many vision statements are overly safe, politicized and uninspiring documents which provide little inspiration for people to innovate. On the other hand, vision statements can also be overly ambitious in seeking to deliver the year-on-year growth which is increasingly demanded by shareholders and markets these days. This can dissuade innovation because many new ideas take longer to materialize than the ever-reducing life-spans of corporate strategies. So an effective vision should be inspiring and stretching but also realistic.

Having created and communicated a clear and realistic mission and vision for a company, the next stage of the traditional strategic planning typically requires some analysis of the forces, risks and opportunities at play in an organization's business environment. According to the theory, this 'strategic analysis' must be done in the context of an organization's unique internal resources, capabilities and stakeholder expectations. The next phase of the process involves formulating strategic objectives and identifying the range of possible actions which will best maximize potential opportunities and mitigate risks in the business environment. Then comes the decision-making phase which involves rationally evaluating and choosing from the alternative courses of action based on their ability to achieve the defined strategic goals. The penultimate phase involves implementing those decisions by acquiring the necessary resources and skills and developing appropriate company structures, systems and cultures. The final phase involves measuring actual performance against strategic goals and taking corrective action by revising goals, strategies or implementation plans where needs be.

The origins of traditional strategic thinking

Cummings' (2002) critical analysis of traditional approaches to strategic management shows that the approach to business strategy described above is rooted in 'modernist' philosophies which promote a highly predictable, deterministic and scientific view of the world. These ideas originated in the seventeenth and eighteenth centuries and are associated with the work of greatest thinkers such as Descartes, Galileo and Newton who questioned, often at great personal cost, the traditional religious and mythological beliefs which dominated thinking at that time. These beliefs were founded on the idea that uncertainty and chaos were an inevitable part of life, that divine fate determined people's destiny and that complete certainty of knowledge was impossible. This meant that new ideas had to be conservative and occur alongside, rather than supplant, the new. In contrast modernist thinkers like Descartes, Galileo and Newton believed that there were fundamental truths to the world and that uncertainty and chaos could be controlled through objective, logical and reasoned thinking (scientific method). These ideas and the subsequent development of science eventually led to the technological developments which formed the foundation of the industrial revolution. And out of this grew new capitalist economies driven by large bureaucracies which required strict rules, processes and controls to ensure predictable and measurable targets against which progress could be measured.

Arguably the most well-known manifestation of modernist ideas in business was Frederick Taylor's (1911) 'Scientific Management'. Adopted by the founding fathers of strategic thinking such as Chandler (1962) and Ansoff (1965), this approach conceived organizations as pyramids with strategists at the top, sending primarily downward flows of directions, instructions and plans to operatives at the bottom. Corporate strategy was strictly the province of the Chief Executive Officer (CEO) and Board of Directors and implementation followed in a linear sequential fashion. Perfect information was also presumed to be available for top management about a relatively stable external environment. And by understanding this environment, it was possible to develop long-range plans based on the extrapolation of future trends. These plans were best developed by specialist planning departments which would then assist the CEO with the detailed task of strategic planning. In essence, strategic management was an exclusive executive function detached from the minutia of day-to-day tasks and differentiated from 'tactical management' (which was about middle managers developing detailed plans to achieve the direction) and 'operational management' (which was about supervisors undertaking day-to-day tasks to put those plans into action).

While Chandler and Ansoff first conceptualized the notion of strategy their ideas largely remained theoretical until Michael Porter (1980, 1985) popularized the concept by developing several simple digestible frameworks that could be used by practising managers to maintain competitive advantage. Porter's three most influential models are; the *value chain*; the *five forces* of industry; and the *generic strategy matrix*. Porter's Value Chain model helps strategists consider how the various parts of a supply chain add value to the end-product or service, thus identifying weak points which need reform. The idea is that firms' activities are undertaken through a supply and demand chain, that value is created for customers in various parts of that chain and that only a few of those chain components are responsible for most of the value. To achieve competitive advantage this is where resources should be focused. Porter's Five Forces model identifies the five main forces which determine a firm's competitive advantage and margins, namely; bargaining power of suppliers, bargaining power of buyers, threat of new entrants, threat of substitute products and rivalry between existing competitors. Competitive advantage and margins are maximized in markets where all of these forces are minimized. Finally, Porter's Generic Strategy Matrix focuses on how firms can differentiate themselves from their competitors. Put simply, firms can develop differentiated competitiveness by adopting three possible strategies: cost leadership (compete on price); differentiation of product or services; or focus on a niche market where it enjoys advantage over competitors.

To Porter, whose work is still highly influential in many industries including construction, the essence of strategy is about making choices and trade-offs and about choosing to be different rather than trying to be all things to all people. Porter argued that good strategy starts with a different value-proposition which creates a 'territory' in which a firm seeks to be unique and dominate its competition. This in essence provides the rationale for investing in innovation since new ideas are the best source of this differentiation. Porter also argued that strategy must have continuity. However, he also recognized that every firm and industry does experience occasional (but rare) 'inflection points' when the needs of customers change to force a change in business strategy. Innovation plays a large part in defining these inflection points which represent opportunities for 'first mover' firms to experience super-profits, at least until their competitors copy and bring the market back to equilibrium. When this happens, and there is nothing to differentiate firms, customers are forced to choose between competing firms on the basis of price only. This in turn encourages companies into price wars and destructive competition which devalues an industry (a scenario which is familiar to many firms in the construction sector). To Porter, forced price competition due to poor differentiation is very different to pursuing a deliberate cost leadership strategy. Given long-standing concerns about destructive price competition and low margins in the construction industry, Porter's ideas suggest that the development of innovative capabilities by firms in the sector has the potential to realize significant competitive advantage.

The limits of rationality

The modernist approach to business strategy has dominated management thinking for five very good reasons (Bungay 2010):

1. It provides comfortable and easy formulas for busy managers to understand and follow.
2. By following established and accepted norms, it is easy for managers to justify and defend their actions.
3. It facilitates the involvement of external specialists who can add credibility to the plan or be blamed if it goes wrong.
4. It lends itself to major and easily coordinated annual planning exercises (often by a separate strategic planning department) which can address the needs of the whole organization for the entire year.
5. It is the default process taught in many management courses and therefore comes with a certain amount of unquestioned legitimacy.

While the application of science to management thinking has undoubtedly unlocked the door to huge increases in business productivity, the predictable, hierarchical world in which these ideas were developed no longer exists. As Bungay (2010) points out, in today's increasingly complex, dynamic and interconnected business world, business strategy cannot be perfectly planned in advance but has to evolve in response to continually changing circumstances. Bungay calls this approach to business strategy 'directed opportunism' and argues that the challenge for contemporary executives is to build an organization which is capable of executing strategy in a fast-changing environment by balancing both high alignment and high autonomy. Metaphorically, this involves seeing a business as an 'organism' which comprises a set of relationships rather than a 'machine' which is constructed from a set of designed parts. In this new model of organization, executives should simply articulate their intent and ensure that juniors understand it, that they have the skills to appraise their own situation and that they are able to make rapid decisions to realize it, often based on partial information.

Bungay's criticisms of traditional strategic management thinking reflect much wider concerns. For example, Cummings (2002) argues that while the scientific and highly prescribed approach to strategy may be useful in simple organizations operating in relatively stable business environments it has considerable limitations outside this context. First, being very time-consuming and resource-intensive, it can reduce an organization's ability to respond quickly to unexpected opportunities. Second, there are also questions about whether those at the top of an organization are best qualified and positioned to detect future trends in the business environment. And third, there have also been concerns that the traditional approach can stifle innovation by 'locking-in' strategies. Given the considerable resources (sunk costs) invested in developing such strategies, managers will be reluctant to adapt and change what may become an outdated and inappropriate plan. Finally, given the magnitude of the strategies that tend to emerge from extensive planning processes, there is a danger that they may be too drastic to accept and never gain widespread support during the implementation phase.

In another critical review of traditional approaches to strategy, Stewart (2009) contends that the term 'strategic planning' has turned out to be an oxymoron. In most companies, he argues, strategy has simply become a ritualistic annual process of following the traditional steps described at the start of this chapter. These models are so ingrained into the minds of most managers that few question their merit or whether there is any evidence to back up their claims of producing better business results. More often than not, this is not the case. For example, Stewart argues that Michael Porter's Five Forces model, which claims that there are three generically successful business strategies, is not based on anything other than anecdote and limited observed experience. One of many untested propositions in Porter's model is that strategy should aim at the singular goal of excess profit through an analysis of industry sectors and the exploitation of their structural barriers and market imperfections. But recent research points to the important interactions between social, environmental and economic goals and has questioned the use of singular metrics in measuring corporate success. In recent years, Porter has responded to such criticisms through the development of the concept of 'shared value' (Porter and Kramer 2006). However, this idea is in its infancy and traditional approaches to business strategy continue to be taught and used throughout the corporate world as true and correct.

A new generation of strategy in construction

As Hamel (2002) points out, nearly everything we know about strategy, organizing, managing and competing has been inherited from an industrial age in which diligence, efficiency, exactitude, quality and control were the complete secrets to success and where industry boundaries were certain, customers loyal and business models eternal. These conditions no longer exist and in response to the above criticisms, thought leaders in the field of business strategy argue that we are moving into a fourth generation of strategic thinking which is more comfortable with eclectic approaches, more politically astute, less dependent on universal solutions and more sensitive to the uniqueness of every organization (Cummings and Dallenbach 2009; Clegg et al. 2011). In other words, contemporary business strategy will need to be based more on judgement rather than prescription.

Loosemore and Phua's (2011) research suggests that this fourth generation approach to strategy is far better suited to the construction sector than traditional approaches. This is because traditional approaches were developed for large organizations which were assumed to:

- operate in stable, controllable and predictable business environments
- have inputs and outputs that were reasonably clear
- have high capital intensity and tightly coupled systems (such as an assembly line)
- have processes which could be broken down into simple mechanistic operations
- operate in an industry sector defined by products and services settled around dominant designs, well-defined and stable business relationships.

These conditions are only found in small sections of the construction industry (such as housing). In contrast, most construction generally occurs within a relatively uncertain and cyclical project-based environment and most firms do not have the luxury of guaranteeing long-term markets for their services. Although there are some very large firms, which are most often subsidiaries of much larger firms, the construction industry is dominated by small-to-medium-sized businesses and few firms, if any, control their environment to the same extent as manufacturing firms can. Furthermore, despite some limited use of automation, robotics and off-site fabrication, construction is not a tightly coupled production line process. Rather, most buildings are largely unique in their design and production environment and production teams are highly dynamic and not based around stable business relationships. Although some standardization of production processes is possible, construction remains a highly labour-intensive and bespoke process compared to the manufacturing sector. This means that inputs and outputs are not always clear at the outset of a project, that most firms are relatively loosely structured compared to manufacturing and that business relationships are generally quite dynamic and changeable.

What do fourth generation businesses look like?

It is all too easy to criticize traditional planned approaches to strategy at a conceptual level but much more difficult to translate these concerns into alternative practical ways of organising. What does a fourth generation business look like and how does it work?

In answering this question, the work of Eisenstat et al. (2001) and Samson (2011) may be useful. Their analysis of leading companies like Citibank, IKEA, 3M, Apple, Proctor and Gamble and IBM shows how businesses can encourage people in lower reaches of their organizations to show initiative and contribute to future strategy. Rather than viewing their businesses as a portfolio of business units, these companies view themselves as a flexible portfolio of resources which they can bring to bear on the most promising opportunities in their sector. In these decentralized opportunity-based firms, quasi-autonomous business units are connected by a corporate centre. Resources from multiple business units are not held in silos but are organized around emerging opportunities in different parts of the organization. Through strong innovation-focused strategy and leadership, opportunity owners are authorized through central screening to mobilize whatever resources they need, allowing these very large firms to emulate the market responsiveness of start-ups while also capitalising on the advantages of their scope and scale. In this way, rather than individual business units pursuing opportunities alone, individual entrepreneurs look for opportunities which cut across different business units and which integrate the business as a whole. This organic structure constantly shifts in response to opportunities and business development managers, who were perhaps once senior project managers, move relatively independently seeking and leveraging new project opportunities by bringing cross-functional teams together from across the business. In effect, these companies have a matrix structure and employ a number of fluid 'opportunity units' supported by a host of stable 'resource units' in order to tackle specific projects.

Many organizations in the construction sector already operate in this way. However, while organising around opportunities in this way has its advantages, it also has its challenges. For example, opportunities must be rigorously screened for strategic relevance, for feasibility and return on investment. This is often a problem in the construction sector because the cyclical nature of the industry makes it hard to plan work and to turn down projects which may maintain cash flows through lean times. Research also shows that there must also be a capability to bring diverse resources together on individual projects which requires flexible and mobile resources which can be shifted easily from project-to-project. Once again, this can be problematic in the construction sector because of poor human resource planning, skills shortages, restrictive trade practices and a lack of multidisciplinary education and multi-skilled training. Another requirement of successful fourth generation organizations is the need for senior managers to maintain open dialogue with lower-level managers. And there is also a delicate balance to keep between managerial control and employee freedom. Too much structure and bureaucracy and the business will grind to a halt. Conversely, too much freedom will ensure that chaos prevails. Furthermore, in contrast to the traditional line-and-staff organizational structure where human resources are connected directly to a central organization for career progression, remuneration, accountability etc., matrix structures require employees to report to multiple leaders, juggle independent assignments and to interact with a wider web of people. As we have seen in the construction sector, there is also a tendency for central business units to casualize their workforce, leading to the 'hollowed-out' firm of individuals with little connection and sense of responsibility to the business. Dainty and Loosemore (2012) argue that this has been responsible for many of the construction sector's enduring problems, such as relatively poor safety, low productivity, high levels of conflict and confrontation and low reliability in terms of delivering projects on time and within budget. As discussed in chapter 3, this represents one of the greatest barriers to innovation in the industry.

Strategy convergence – the catalyst for innovation

Hamel (2002) recognized the limitations of traditional approaches to strategy. In particular, by 'locking-in' plans, a firm's strategy will inevitably decay over time, eroding a firm's capacity to differentiate itself from its competitors. In the long term, as firms fail to differentiate themselves, there is a tendency for firms in most industries to cluster around some central tendency of industry orthodoxy. This is a well-known phenomenon in economics known as institutional or industry isomorphism (DiMaggio and Powell 1983) and is based on observations that over time, firms in any industry tend to exhibit common patterns of behaviour. This in turn drives competition towards margin-based strategies, since the prices of raw materials and labour are also likely to converge.

Hamel (2002) argues that it is only through innovation that firms can escape this profit-eroding cycle of 'strategy convergence'. However, innovation is not a one-off activity and needs to be continuous to maintain competitive advantage. While successful entrepreneurs can dominate a market and make super-profits in the short term, their excess rewards are ultimately fated to be eroded by the inevitable appearance of imitators and rivals who are eager to share in the bounty. Although the speed of this 'strategy decay' which is the fundamental cause of industry homogeneity depends on competitor agility and market barriers to entry, this ultimately forces the entrepreneur back to the drawing-board to search for innovations which can generate another round of above normal earnings.

To lead or follow?

Industry homogeneity represents enormous opportunities for firms that can first recognize the signs of strategy decay and innovate to differentiate themselves from their competitors. But the decision to break with the pack and take an industry lead in grasping these opportunities is not as straightforward as it may seem. As Lieberman and Montgomery (1998a, 1998b) point out, the main advantage afforded to first movers is early revenue growth, often at super-profit levels as they dominate a market. These super-profits arise because first movers are able to benefit from technology leadership by developing their understanding of and use of technology in ways that are hard for later entrants to copy. First movers can also establish their product as the industry standard, making it more difficult for followers to gain customer acceptance. Another benefit realized by first movers is the opportunity to control resources that later entrants must use to compete. For example, first movers into sustainable development have been able to secure alliances and partnerships with the best researchers, suppliers of sustainable raw materials and service providers. They can also benefit by securing patents which prevent other companies from copying their ideas. A particularly strong advantage arises when the first mover can build resources that may discourage entry by other companies. For example, a first mover can increase production capacity or broaden their product line, signalling to competitors that there is not enough room for followers to enter and profit. First movers may also enjoy the advantage of buyer-switching costs. This occurs when it is costly or inconvenient for a customer to switch to a competitor once a first mover has secured their custom. Examples of switching costs include: adapting production processes to use new raw materials; training employees to work with new partners and technologies; potential risks associated with switching the unknown products and services; and penalties associated with breaking a long-term contract. In this way, the first mover has the opportunity to shape consumer preferences and build brand loyalty.

Research has sought to understand the conditions where first-mover advantages are most likely to occur (Sharad and Swami 2004). For example, it has been found that consumer product markets appear to offer more first-mover advantages than one-off batch production markets like construction. This is because first movers can market their new ideas or products to large volumes of consumers and lock themselves into a continuous revenue stream to capitalize quickly on any advantages they can secure. In contrast, most firms operating in the construction sector are not dealing with a mass market and have to wait for opportunities to come to them. Research has also showed that that first mover advantage is higher when switching costs are higher. However, in general, buyer-switching costs are not likely to be high in construction since it is a highly competitive industry with few barriers to entry and there is little to differentiate one firm from the next. Furthermore, the one-off nature of the product makes it harder for firms to seek exclusive rights to work in an ongoing relationship with particular clients. A final important factor in securing first mover advantage appears to be the ability of the first-mover firm to mobilize its resources to secure its initial advantage. Firms that can easily and quickly switch resources from one area to the next are able to capitalize more from a new idea than those which move slowly. As discussed above, this can be problematic in the construction sector because of poor human resource planning, skills shortages, restrictive trade practices and a lack of multidisciplinary education and multi-skilled training which reduces workforce flexibility.

Taking into account the above discussion, company strategists need to decide if they are likely to benefit from being a first mover, or whether it would be better to wait and follow. As Kerin and Cook (1989) points out, while moving fast can provide many competitive advantages, moving too fast can lead to costly mistakes and failures. Moving ahead of a market can mean that the demand for the innovation will not exist, no matter how good the new idea is. Clients and customers therefore play a critical role in determining the appropriate timing of the innovation process and this is especially so in the construction sector where the ratio of clients to products is far smaller than many other sectors such as manufacturing. As Loosemore and Phua (2011) found in their research into the sustainability strategies of Australia's leading construction firms, clients do not always see the benefits to their core business of investing in innovation. While there are clearly exceptions to the rule, many construction clients are driven by minimizing initial capital costs rather than maximizing long-term value over the life-cycle of the built assets they procure.

In this risk-averse environment, careful incremental movements may be wiser than sudden leaps forward. For example, researchers have found that in most industries it is relatively easy for later entrants to learn new innovations quickly (without the early failures and risks) and overcome the lead held by the first-mover firm (Lieberman and Montgomery 1998a, 1998b). Also, in some situations later entrants have been able to adapt and reinvent new innovations quickly enough to void a first-mover's patent protections. In most industries patents confer only weak protection, are easy to circumvent or have transitory value given the pace of innovation and technological change. Indeed, sometimes companies that enter a market later can achieve superior results to those realized by the first-mover firm and capture a greater share of the market. For example, the first entrant may invest heavily in enticing customers to try a new type of product meaning later entrants can benefit from informed buyers without having to spend as much on marketing and education. As Spence (1981) points out, first movers will often need to sell their products below cost in an effort to understand the market better. Furthermore, follower companies can reverse-engineer many new products to develop superior products either faster or cheaper, negating much of the first-mover advantage. In other words, later entrants have the ability to study the first mover's mistakes, techniques and strategies and 'free-ride' on a pioneering firm's investments without incurring the significant up-front costs which the first mover has to sustain to create a market from scratch. This has been vividly illustrated over the last thirty years with the economic emergence of developing countries like Japan in the 1970s and more recently China, which have grown on the back of imitating and improving existing Western products. While we tend to denigrate adaptation as an inferior form of innovation, there is nothing inherently inferior about the process of imitating and adapting another's ideas. Indeed, one could argue that every innovation has some elements of borrowing from other innovations. For example, Arora et al.'s (2001) research into the innovative activities of 133 of the world's most innovative firms like IBM, AT&T, Motorola and Microsoft found that 77 per cent had licenced technologies from others and that 62 per cent had licenced technologies to others. As we have argued in previous sections, many leading firms now innovate by forging inter-firm alliances and patent pools that enable technology corporation and cross-fertilization and sharing of new of ideas.

It is clear from this discussion that companies need to carefully consider the relative benefits of being a leader or follower. Questions they may ask might include:

- How big is the opportunity if we move first?
- Can we mobilize our resources quickly to take advantage of the opportunity?
- How much up-front investment is needed?

- Do we have the resources to invest?
- Are our customers and clients ready to support such a move?
- Will they need educating or will we lose them?
- How great are our customers' switching costs (can we lock customers in)?
- What are the opportunities of being first in a market relative to the risks?
- What are the opportunity costs of moving first?
- If we move first, can our competitors secure second-mover advantage?
- Do we have the resources to sustain our advantage?
- If someone else enters first, how difficult will it be to follow and what disadvantages and advantages might later entry provide?

Clearly, some firms are more suited to be pioneers than others. This depends on their:

- appetite for risk
- back-up resources to cover inevitable failures and uncertainties
- effective risk and opportunity management capability
- loyal and supportive customers with a long-term interest in their services and products
- strong leadership, strength of conviction and a clear vision
- strong innovative cultures, values and beliefs to provide the determination and climate to succeed through the inevitable ups and downs.

So it is clear that one cannot generalize about the advantages of first-mover status in the construction sector. The pioneer strategy is not necessarily a route that just any construction firm can take but in the right circumstances it can result in significant returns. However, managers must remember that first-mover advantages are not everlasting and that innovation must be a continuous process to maintain a competitive edge. Eventually, one thing is guaranteed. The competition will manage to take at least some piece of the market and firms must put in place strategies to prevent complacency, maintain innovation and avoid inertia caused by insensitivity to future changes in the market environment.

Enlightened business strategy

We have argued that traditional approaches to innovation are inadequate for three main reasons:

1. They dictate over-simplified, standardized and rigid approaches which fail to recognize the uniqueness of every organization.
2. They are grounded in deterministic 'modernist' management philosophies which fail to recognize the dynamic, interconnected and uncertain world in which we live.
3. They undervalue the importance of people and stakeholder rights in a more informed and socially conscious world.

Our discussions have led us to the conclusion that an alternative approach is needed and in the following sections we identify what the essential components of a more enlightened innovation strategy should be.

Utilizing corporate identity

Cummings (2005) argues that in an increasingly homogenized business world where it is difficult to differentiate between many firms, innovation depends more than ever on the one organizational resource that other firms cannot copy (its corporate identity). The concept of corporate identity is akin to a firm's unique spirit, character and personality and is rooted in its unique relationships, culture and history. Sometimes referred to as a business's DNA, a firm's corporate identity is defined by its stories, heroes and outlaws, artefacts, symbols, language, rituals, rules and traditions (Driskill and Brenton 2011). Many companies neglect to leverage this critical asset because it is tied up in a firm's tacit knowledge. As Nonaka and Takeuchi (1995) found in their influential analysis of innovation in Japanese industry, tacit knowledge is difficult to use because it is under the surface and implied rather than explicit. This makes it very difficult to 'get at' and once lost, it is very difficult to get back.

Authors like Cummings (2005) and Swann et al. (2004) point to accumulating evidence that the world's most innovative companies use their corporate identities to inspire strategic differentiation. The advantage of this approach is that it not only inspires difference but it builds legitimacy, authenticity and loyalty with clients because strategic initiatives are seen to be aligned with corporate values and to be sustainable in the long term. Firms that use their corporate identity to differentiate themselves are also more memorable than those which do not. Organizations such as Virgin, The Body Shop, Google, Apple, Salvation Army and Coca Cola have been highly effective at leveraging their corporate identity to build highly memorable brands to dominate their respective sectors. The one common characteristic that links these very different firms is that we all know 'what they stand for'. To what extent can the same be said for firms in the construction industry?

The creation and maintenance of a clear corporate identity is the responsibility of senior managers whose primary role as leaders is to create a unified identity that organizational members can understand, relate to and follow. The answer to the question 'who are we?' will determine not only how people behave, but how they interpret the future, identify threats and opportunities, communicate internally and externally and resolve conflicts. Beliefs about identity also affect the ways leaders craft strategy, the types of goals they pursue, the types of customers and business partners they work with, and how they allocate and expend resources. Finally, when there is a clear corporate identity, people are able to 'identify' more strongly with an organization which in turn leads to higher organizational commitment, engagement, acceptance of change and cooperation among employees.

As Swann et al. (2004: 10) note, organizations without a clear corporate identity become a 'fertile breeding ground for misunderstanding and discord'. Here leaders find it difficult to build momentum for new projects and initiatives. And disagreements about corporate identity may result in stakeholder confusion or even client mistrust because of contradictory external messages. So consensus about identity is one of the cornerstones of innovation. The more time companies have invested in devising strategy in ignorance of their own corporate identity, the more like their competitors they will become.

It must be noted that while a strong and clear corporate identity appears to be important in driving innovation, the jury is still out on the exact nature of this relationship. For example, recent research indicates that some degree of ambidexterity may benefit organizations (He and Wong 2004). It also recognizes that multiple identities (for example in different geographical regions) are inevitable and necessary and might actually foster innovation rather than stifle it (Foreman and Whetten 2002). The basis of this argument rests in the diversity of perspectives which multiple identities can provide. Also, while multiple corporate identities can cause confusion, it can generate creative tensions, afford flexibility and provide a wider vision in identifying and responding to business trends. A third possible benefit is that organizations may have the capacity to meet a wider range of stakeholder demands (Weick 1976).

Taking an emergent view

We live in the 'information age'. Over the last two decades, the ephemeral and increasingly shared nature of communication technologies has fundamentally changed the fluidity of knowledge and the rate at which it becomes available and obsolete. It has also changed traditional processes and timeframes by which knowledge is acquired and collectively understood. As Cormier (2011) points out, as the pace of change in information technologies and the rate of connectivity in society have increased, so has the currency and transience of what we traditionally know as knowledge. What counts as accurate, current, past or future knowledge has become unclear and the creators of new knowledge are no longer easily identifiable as traditional formal and verified experts. Instead, knowledge is now generated by a wide array of 'knowers' sharing in the collective construction of a field of knowledge. In other words, today knowledge is increasingly socially constructed and knowledge acquisition becomes a negotiation between various sources rather than a taken fact, as it might have been a decade ago.

To deal with this world of increasingly transient knowledge, a more 'emergent' approach to business strategy is necessary. Indeed, it has been known for some time that successful firms have both formal/planned and informal/fuzzy elements to their business strategies. According to both Minztberg (1984) and Hubbard et al. (2002), despite what the strategic management literature says, successful firms do not stick slavishly to what they said they were going to do. Instead, they recognize that the future cannot be planned in infinite detail and that they need to be constantly sensitive to new knowledge and able to respond quickly to convert it into business opportunities. It is interesting to note that Hillebrandt and Cannon (1994) found that the strategies of construction firms have a large emergent dimension. Managing a construction company successfully involves a constant opportunistic juggle to match human resources and management skills to a constantly changing array of geographically dispersed projects over time. More recently this was supported by Green et al. (2008) who found that in construction firms, strategy is more often emergent than pre-planned, shaped by unexpected opportunities and by individuals and often maverick behaviour. Green et al. found that while boards may intervene with planned strategies occasionally, there was little evidence that formal strategic planning techniques were used or that they had much impact on enacted strategy in practice. This suggests that the strategic flexibility to innovate exists in the construction sector. However, rather than being driven by the need to innovate, this would appear to be a function of the industry's need to organize within an uncertain project-based structure. The challenge for firms in the construction sector in the future is to put this natural advantage to good use in driving innovation across their business.

Being inclusive

Pettigrew et al. (2006) and Werbac (2009) recognize that one of the main limitations of traditional models of strategy is that they neglect the critical role of social engagement and stakeholder interactions in the strategic decision-making process. They argue that strategy should be more inclusive than the top-down process dictated by modernist approaches. It is also increasingly accepted that although commercial firms have a profit objective, they also have to pursue other social and ecological objectives which may lead to trade-offs between long-term and short-term performance. Furthermore, senior managers are not as rational and logical as traditional approaches suggest. Instead, in analysing and responding to the future, managers appear to be influenced heavily by political forces, are more closely connected to lower operational workers than has been assumed and are not as aware of their business environment as implied by traditional models of strategy. Indeed, lower level managers are often more in touch with the business environment than senior managers, affording them access to important ideas and knowledge about the future which top managers do not have. In summary, contemporary strategic thinking sees innovation as an inclusive process which facilitates the opportunistic synthesis of ideas from all organizational levels.

Developing unique resources

The resource-based view (RBV) of the firm recognizes that every firm has a unique pool of resources which are greater than the sum of its parts. These resources include:

- physical/technological capital (land, premises, plant etc.)
- financial capital (cash, capital etc.)
- human capital (people, experience and expertise etc.)
- social capital (relationships, networks and connections)
- intellectual capital (knowledge and ideas)
- organizational capital (culture, reputation, relations etc.).

The RBV argues that in every firm these resources interact in unique ways to give a firm its distinct identity. It also argues that a firm focuses on creating competitive advantage in the market by building unique 'inimitable capabilities' and 'core competencies' based on its unique resources (Prahalad and Hamel 1990). A firm secures competitive advantage in the long term through 'dynamic capabilities' in managerial and organizational processes which allow it to innovate to adjust these core competencies and resources to market changes (Teece et al. 1997; Eisenhardt and Martin 2000).

The value of the RBV in construction was recognized by Barrett et al. (2008). To them the potential for innovation does not come from the unique resources a firm owns but from how a firm uses and develops them to drive innovation. Barrett et al. (2008) argue that the proactive 'resource-push' view of innovation offered by the RBV (where firms innovate because they can) provides a more stable grounding for innovation in the construction sector than the dominant reactive 'market-pull' orientation (where firms innovate because they are asked by clients). As we shall see in the next chapter, the construction sector cannot rely on its clients to drive innovation and must take the responsibility to do so itself.

Although many types of resources contribute to an organization's capacity for innovation, Dodgson and Gann (2010) point to research which indicates that an organization's most important resources for driving innovation are (in rank order):

1. The energy, imagination and knowledge of its employees. This resource need to be harnessed and supported by stimulating workplaces and by flexible organizational structures, systems and practices.
2. A firm's external social capital which resides in its relationships with customers, clients and suppliers of goods and services.
3. A firm's sources of intellectual capital which in order of priority include professional conferences, academic and trade journals and trade fairs, followed by universities and government research laboratories.

However, as Dodgson and Gann point out, simplistic research which seeks to separate out the drivers of innovation in such a clinical way disguises the underlying complexity of innovation processes. For example, the placing of universities and government research institutions as being relatively unimportant to innovation reflects a particular 'industry' perspective. It also reflects the short-term perspective of many managers and widespread ignorance of the critical importance of research in seeding long-term advances in knowledge and innovations. As we show in the next chapter, the gestation period between the discovery of a new idea by a researcher and its uptake by industry can take many years or even decades and often obscures where that idea originated. In reality, innovation normally involves a combination of interdependent ideas generated by a multitude of resources over long periods of time which come together in complex and sometimes serendipitous ways within different social, political and economic environments and constraints.

In addition to helping understand the role of corporate resources in the innovation process, the RBV also tells us that a firm's ability to innovate is constrained by 'path dependencies' dictated by the traditions and history of a firm's culture, policies and resources. So for example, older construction firms with long established traditions and cultures are likely to find it more difficult to innovate in response to changing external factors than younger firms. Green et al. (2008) argue that the existing literature in construction says very little about path dependencies and rarely are firms encouraged to think about how to adapt their organizational processes to changing circumstances. Instead, according to Green et al. (2008: 65), construction firms are uncritically 'exhorted to adopt supposed universal best practice recipes such as lean thinking, partnering and integrated teams' and the 'overriding tendency is to conceptualize managerial skills as a static resource which is acquired and deployed as necessary' rather than developed and adapted to changing circumstances.

Collaboration

According to Manuel Castells (1996), one of the most defining transformations in society and business over the last forty years has been the growth of the 'networked society'. This has been characterized by five prominent trends in the business environment:

1. The emergence of new information technologies
2. The growth of globalization
3. The replacement of vertically integrated businesses with loosely coupled networked organizations
4. The growing popularity of outsourcing; growing workforce casualization
5. Greater potential for contagion between different global and sector-specific markets.

Castells argues that these trends emerged during the 1970s as new flexible forms of organization emerged to deal with the growing economic uncertainties associated with increasing liberalization and globalization of trade. It also became evident for the first time that the planet had limited resources and that these were controlled by a small number of relatively unstable countries. Before the 1970s the main source of competitive advantage was economies of scale and scope of production. But after the 1970s, as a result of these fundamental changes in the business environment, strategic advantage relied more on an ability to innovate as a means of differentiating oneself from potential competitors. As Friedman (2005) points out, to compete effectively firms needed to identify and develop their core competencies, building up world-class knowledge in a few restricted areas, a trend which inevitably required them to collaborate in inter-firm networks of complementary skills. With the continued growth of globalization and the development of even more advanced information technologies the concept of the 'networked society' has become reality.

In this new networked business environment, most innovations now arise from collaborations 'between' different companies rather from 'within' one company, as was the case in the past. Recent research shows that firms which strategically build and leverage their external business networks are significantly more likely to innovate than those which do not. According to Horrigan (2011), in the future organizations will increasingly work in 'swarms' – a form of organizing characterized by flurries of collective innovative activity where *anyone* is able to add value. These corporate swarms will form quickly, attack a problem or opportunity and then quickly disband. In contrast to traditional notions of team working where people may have had long-standing and stable relationships over time, swarms will be characterized by loose connections and weak ties. And rather than being driven by business, they will be formed by individuals using social media to create personal social networks and will be characterized by a certain degree of unpredictability which will be outside a manager's control. Indeed, these new business networks do not need to physically exist but can be virtual organizations enacting their activities and culture in cyber space.

Smart managers will learn to use these constantly shifting networks to leverage opportunity. They will realize that they now live in a hyper connected business ecosystem which they cannot control but only influence. In contrast, managers who continue to assume the world is stable and to assume they can perfectly control the way that people interact will find the world a very frustrating and dysfunctional place. As Dodgson et al. (2005) point out, this 'open' and 'distributed' model of innovation is very different to the closed, individualized and corporatized models of innovation which dominated thinking in the past. It means that firms have to clarify and establish their position in the value stream much more precisely than before. Furthermore it means that they will have to focus on building unique collaborative business networks in their supply and demand chains through partnerships, alliances and joint venture arrangements etc.

These profound changes have been evident in the construction industry. As Langford and Male (2001) pointed out, the most important asset which determines the competiveness of a construction firm is no longer its access to cheap materials or labour but the nature of its relationships within its supply and demand chains. For example, Toole's (1998) analysis of home builders' adoption of technological innovations found that innovation was influenced strongly by their social networks (business partners, competitors and subcontractors and suppliers) and by their connectivity to trusted sources of information. However, research also indicates that the level of sophistication in managing construction supply chains has lagged behind other sectors which have developed in 'deeply' integrated supply chain networks. For example, Tennant and Fernie's (2011) research found that, while over the last decade principles of supply chain management have been promoted extensively within the construction sector, to date the focus has been on tier-one and tier-two relationships and has rarely extended into downstream supply chain relationships, leaving the aim of fully integrated supply chains an aspiration rather than a reality.

In exploring the challenges of developing and managing these new forms of inter-firm networked organizations, de Man et al. (2008) and Driskill and Brenton (2011) found that there are five main problems to overcome.

1. Motivating partners to share knowledge – since the complexity and multilayered nature of inter-firm networks makes the pay-offs for collaborating obscure and often very indirect.
2. Preventing free-riding behaviour – particularly in large networks where it becomes easier for parties to profit without making an appropriate contribution.
3. Realizing efficient knowledge transfer – particularly in large networks where all partners cannot possibly be directly connected to each other.
4. Integration – inter-firm collaboration requires the integration of many organizational procedural, logistical, cultural, syntactic, semantic, time, physical and geographical boundaries.
5. Building a corporate culture – which gives the network a sense of identity and cohesion and basis for acting together as a team.

Research shows that many of these problems can be minimized by:

- choosing partners who are willing to share information
- providing incentives to collaborate
- rationalizing supply chains to build stronger relationships between fewer partners than having amorphous and dispersed relationships with a large number of partners
- training and education to enable firms and people to work across new boundaries
- communicating the clear mutual advantage to participants of long-term collaboration rather than competition
- resolving contractual issues to ensure that firms take a longer-term view of collaboration and that there are clear contractual rules of IP ownership and value appropriation
- resolving communication problems by having effective knowledge management systems to facilitate the transfer of information, by maximizing opportunities for people to meet (through rotation programmes, co-location and communities of practice etc.) and by making maximum use of

information technologies to lower the search costs associated with finding information.

While external collaborative networks which integrate supply and demand chains are critical to driving innovation, so are internal collaborative networks within firms. Bryan et al.'s (2007) research into informal networks in large organizations like ExxonMobil and GM showed that firms have dozens if not hundreds of informal networks which rarely reflect dictated communication routes in formal organizational charts but which form and shift constantly as a result of the voluntarily activities of employees. Few managers understand how to harness the power in these networks and they usually form unpredictably around new information technologies such as mobile phones, community web sites and social media. So far we have argued that the power of such networks is that they cut across organizational silos, enabling serendipitous insights and collisions which often lead to new innovations and discoveries. However, Bryan et al. argue that while serendipity can lead to unexpected opportunities, there is also the need for some element of control. In reality few managers are able or willing to expose their businesses to the chaos (albeit potentially creative) that serendipity brings. Informal networks when left entirely to themselves can become highly dysfunctional presenting new complexities, muddling roles, intensifying politics and presenting invisible risks to an organization. The compromise which Bryan et al. propose involves creating 'collaborative networks' which Bryan et al. point out should not be confused with matrix structures.

Matrix structures work through formal authority and maintain organizational hierarchies. In a matrix structure, employees have two or more bosses who have authority over their work and the person on the power axis must resolve any conflicts in direction and resourcing which might arise. Bryan et al. argue that while this structure loosened up organizations enough to cope with the business world of the 1980s, they are not flexible enough to enable organizations to harness the opportunities presented in a more uncertain, dynamic and interconnected world. The increased opportunity for informal interactions between employees bypasses formal matrix structures and confuses accountability for decisions. Furthermore, the vast amounts of information which are now available to decision-makers swamp matrix managers creating symptoms such as endless meetings and inertia from paralyzed decision-making. Consequently, employees who wish to work horizontally across organizational boundaries find themselves forced to work through established vertical and horizontal silos which the matrix structure preserves.

In contrast 'collaborative networks' organize through mutual self-interest and collaboration. They have a defined leader (best selected by the members) and are focused around specific topics which are of strategic importance to the business. The leaders of these groups are not 'bosses' but are 'servant leaders' who help the group achieve its aim by facilitating an agenda, inspiring and assisting members to collaborate, creating a vibrant network culture, organizing infrastructure and securing resources. While this idea of formalizing an informal network appears counter-intuitive, these collaborative networks retain some element of informality since they can form and disband at any time, members can come and go as they please, membership is entirely voluntary and so are their rules of engagement. In effect they are self-organising communities of practice rather than formal groups. However, managers retain some element of control and assist in the informal interactions by enabling an infrastructure and resource base to support and facilitate the ongoing exchange of ideas. Furthermore, by giving them some form of leadership, they prevent the type of destructive structures emerging where certain individuals can act as arbitrary gatekeepers to membership and hoard or control information flows to their personal advantage. Organizations might even help encourage membership through incentives to engage and shared rewards for any group-based outputs which result. The number of collaborative groups an employee participates in is normally entirely up to them, although there may be a minimum requirement to participate in at least one group. By participating in more than one group employees are more able to integrate knowledge from different parts of the business. Rather than rewards being based on individual achievement as they are in many organizations, rewards should be based on the effectiveness of the group itself which may be measured in both qualitative and quantitative ways (size of group, engagement with group, number of internal interactions, connectedness density, ideas produced etc.).

Leveraging social capital

The changes discussed above, away from vertically integrated organizational structures to new horizontally dispersed network structures, are challenging the myth of corporate individualism that has guided traditional approaches to innovation and strategy for so long. Instead of trying to be everything to everyone, successful firms are now competing in restricted market segments and layers while promoting strategic alliances with players in adjacent layers and other industries to create a unique cocktail of knowledge that will deliver a competitive edge. So innovation and corporate success are no longer an individual matter. Rather, it depends on the quality of corporate relationships with other firms and the value that resides in these relationships (a firm's 'social capital'). Social capital refers to the value imbedded in a firm's social networks which is dependent on the power and connectivity of people in those networks and the firm's position within them (Burt 2005). The concept of social capital is relatively new and is based on the observation that well-connected actors in a social network tend to do better than those who are not so well connected. The management of social capital requires new skills which are very different to those required to manage the traditional sources of capital cited in management theory (people, technology and money). Social capital differs from traditional sources of capital in the sense that it is intangible and tacit, it is located outside the business in relations with others and is not owned by any specific firm. It is also important to realize that a firm's social network does not translate automatically into social capital. Rather, it is the positioning of a firm in its social network and the nature and quality of its relationships with others in that network which create its social capital.

Research in the field of social network analysis specializes in understanding what types of positions and structures afford most power to a firm or individual. It has shown conclusively that if one is connected to people with special resources and powerful connections then one's social capital is also increased (Brass 2003). This research has also shown that certain network positions, particularly those who are 'central' or located 'between' others, derive the most power from their social networks. More specifically, it is those who occupy connecting positions between actors who are disconnected, but who need to interact, who have the most power. Burt's (2005) research suggests that the 'brokers', who fill these 'structural holes', perform particularly important roles in driving innovation. They do this by bringing people together from different knowledge domains in new non-traditional combinations who may otherwise be separated by organizational silos and suspicious external stakeholder relationships. Dodgson et al. (2005) also recognize the importance of brokers in the innovation process by referring to the critical importance of 'T-shaped' people in organizations. These people are critical to innovation because they facilitate the co-creation of new ideas through their in-depth knowledge of specialist areas and their ability to integrate it by brokering connections between different business units. Later in this book we show how brokering alone is not enough and that to function effectively, these relationships need certain attributes such as trust, diversity and equality. But in purely structural terms, innovation can be seen as a process of looking for missing, value-adding connections within and between organizations and brokering them to create new value.

Conclusion

The premise of this book is that as the world becomes more uncertain, dynamic and interconnected, it is becoming more difficult for businesses to plan their way to success. We have argued that many organizations have been caught 'wrong footed' by these changes and rely on outdated management strategies which prevent them from innovating. In this chapter we have shown that innovation requires a flexible business strategy which not only provides agility but which enables the integration of different types of knowledge in complex and dynamic networks of individuals, firms, communities and governments. This collaborative vision of a diverse learning community represents the foundations of an effective innovation strategy and in the next chapter we delve into organizational, sociological, psychological and behavioural research to explore new practical strategies which firms can use to build such a capability.

6 Organizing for innovation

> Somewhere out there is a bullet with your company's name on it. Somewhere out there is a competitor, unborn and unknown, that will render your strategy obsolete. You can't dodge the bullet – you're going to have to shoot first. You're going to have to out-innovate the innovators.
>
> Gary Hamel

Introduction

We have argued throughout this book that there is no single recipe for innovation. Each firm must develop its own unique approach in response to its own culture, opportunities, constraints and capabilities. However, over fifty years of research in the field of management and in the social and behavioural sciences has shown that there are some common attributes which seem to define the shape of innovative firms. The aim of this chapter is to sift this vast body of knowledge to present digestible, jargon-free and practical insights which managers in the construction sector can use to better drive innovation in their business.

Organizational structure and innovation

Over fifty years ago, Burns and Stalker's (1961) formative study of Scottish electronics firms concluded that companies facing a dynamic and uncertain business environment were best served by an 'organic' organizational structure which allowed them to quickly adapt to unexpected opportunities and threats. Organic organizations had a network structure with a central hub. These organizations had little hierarchy, few rigid rules and procedures, placed a high value on external knowledge and were driven by the power of personalities and relationships. At the opposite end of the spectrum were 'mechanistic' organizations where decisions were made centrally. In contrast, companies operating in a certain and stable environment would benefit from maintaining a 'mechanistic' organizational structure. They were rigid and hierarchical, their policies and tasks stayed the same for long periods of time and decision-making was

Table 6.1 Characteristics of organic and mechanistic organizations

Mechanistic organization	*Organic organization*
Centralized decision-making.	Decentralized decision-making.
Hierarchical structure with stable divisions/departments based strongly around functions.	Flat structure with temporary work groups/teams based around specific projects.
Vertical communications dominate.	Lateral communications dominate.
Formal communications dominate.	Informal communications permitted.
Rigid job definitions set by senior managers resulting in a high level of specialization.	Flexible job definitions defined by individuals through interaction and negotiation with colleagues who decide the best way to get the job done.
Many rigid rules and procedures which result in prescribed work patterns.	Few rigid rules and procedures.
Power and authority based upon formal seniority within hierarchy.	Power and authority changing with changing circumstances and based upon individual skills and abilities.

(Adapted from Burns and Stalker 1961)

centralized around a few senior individuals. The contrasting characteristics of organic and mechanistic organizational structures are summarized in Table 6.1.

In theory, organic-type structures provide people with a relatively high degree of autonomy and control over their jobs, giving them the confidence to experiment with new ideas and to interact with others in their environment, without fear of failure. The natural conclusion is that firms with organic structures would be more innovative and perform better in today's uncertain, dynamic and interconnected business environment. However, while this may be true, there are always exceptions to the rule and sweeping generalizations should be avoided. For example, soon after Burns and Stalker's formative work, Woodward's (1965) well-known research found that the relationship between the type of structure and the level of performance was dependent on the type of 'technology' used in their production processes. Woodward found that successful firms associated with 'small-batch' production industries which produced one-off products (such as art works and buildings) were characterized by organic structures. In contrast, successful firms in large-batch/mass production industries which produce identical products (like manufacturing) reflected mechanistic structures. Finally, successful companies in continuous process production which require minimal human intervention (such as chemical plants or nuclear power stations) tended to adopt a hybrid organic structure. More recent research in the field of innovation has shown that while organic structures are best for the up-front development of new ideas, mechanistic structures are best during the implementation phase of the innovation process, because new ideas often require backing from sources of authority to overcome resistance (King 1992).

While Woodward's (1965) analysis posed as many questions as it answered and was limited by its relatively small sample of one hundred companies, it was highly important in challenging, for the first time, the existence of universal principles of management – that there was one best way to manage in all circumstances. The lesson here for contemporary managers is that they should be wary of anyone who is selling universal truths. And they should be very careful not to automatically follow the latest business trends. Few, if any, of these new ideas are innovative. And when one digs beneath the surface, most are quickly revealed as a rebadging of old defunct ideas that have lost their shine. There are certainly no magic bullets to the challenge of innovation. If there is one universal truth, then it is to develop an approach to organizing which considers a firm's unique identity, culture and context. The danger in today's fast moving and highly visible business world is that this basic principle is forgotten in the headlong rush to be seen to be innovating, regardless of whether it provides any benefits.

Systemizing innovation

Samson's (2011) cross-sector analysis of highly innovative firms like IKEA, Hewlett Packard (HP), IBM, Proctor and Gamble, Google, Cisco, Tata, Apple, Sony, Samsung, Toyota and Disney found that they are differentiated by a robust and 'systematic' innovation capability. According to Samson, the secret to 'systemizing innovation' lies in securing six main principles of organization:

1. Customer-focus
2. Collectively challenging orthodoxies
3. Dedicated resources
4. Measuring innovation return on investment
5. Recognizing and rewarding innovation
6. Accountability for innovation.

Customer focus

Innovative firms tend to articulate a clear direction and inspirational vision of creating 'new value' for customers. While senior managers drive the strategy personally and play a pivotal leadership role in modelling innovation, effective innovation is also 'needs-driven' and 'customer-focused'. For example, in companies like 3M feedback from lead users who are interested in trying new products and services are often the sources of new ideas. In contrast, companies with a poor record of innovation tend to have 'competitor focused' strategies – using them as a benchmark in strategy formulation.

Collectively challenging orthodoxies

Innovative firms develop both formal and informal processes which encourage people to challenge orthodoxies and put new ideas forward. For example IKEA challenged existing orthodoxies when it questioned whether furniture needed to be delivered already assembled. So did Toyota when it developed the Prius – the world's first hybrid mass-market car. In highly innovative firms, innovation has less to do with individual creativity than with assembling the right sorts of insights to provoke new breakthroughs. What defines these firms is their ability to combine knowledge from various domains to view things from new perspectives.

Dedicated resources

Innovative firms commit ongoing and substantial budgetary resources to key innovation activities such as R&D, training and development, environmental scanning, marketing and user interfacing. They are also adept at securing external resources for innovation. For example Cisco and IBM wait for venture capitalists to capitalize companies with promising technologies and then, when successful, purchase their innovations through acquisitions. Other firms like Tata in India provide resources for promising ideas through competitions and grants. In 2011, Tata's innovation competition received over 1700 entries from its seventeen subsidiaries.

Measuring return on investment

The returns from innovation activity can be complex and difficult to measure, especially in the short term. Nevertheless, most innovative companies attempt to do so. In general innovation is measured in combination of three ways:

1. Inputs measures (e.g. ratio of R&D to total staff numbers)
2. Innovation process intensity (quantity and quality of innovation activities)
3. Innovation outputs (revenue generated, new processes, products and services, patents, reputation etc.).

Recognizing and rewarding innovation

Innovative firms ted to have developed reward systems that link rewards to long-term measures of success. The short-term measures of success used by many other firms tend to discourage risk taking and be intolerant of the early failures that define many good ideas and they also tend to deter investment in longer-term blue-sky initiatives. The linking of rewards to long-term performance tends to be an attribute of Asian companies. For example, Japanese companies typically have relatively low fixed wages with a large bonus linked to long-term overall firm performance whereas American companies tend to have high fixed wages which are linked to short-term individual performance.

Accountability for innovation

Innovative companies tend to have strong corporate cultures which encourage experimentation and risk taking but which also tolerate change and failure. For example, in HP a strong culture of innovation has been developed through senior managers who have access to a central pool of resources dedicated to the creation of new ideas and are expected to personally drive innovation through the business. While these managers have to report the effectiveness of their activities against well-defined short-term and long-term innovation goals, there is also tolerance to the fact that not all ideas will produce immediate returns and might require repeated injections of resources.

Building a culture of innovation

Just like every country or nation has its own unique culture, so does every organization. According to Deal and Kennedy (1982) who first introduced the notion of an organizational culture, every organization can be defined by its own unique:

- communication styles
- language
- values
- history
- stories
- heroes
- outlaws
- metaphors
- shared symbols
- rituals
- ceremonies
- norms
- artifacts such as buildings, décor, workplaces, logos and codes of dress.

Deal and Kennedy argued that these characteristics define an organization and bind it together, creating a corporate ethos (personality) which is accepted by its members and which is preserved and transmitted as a corporate identity between generations of workers through the process of socialization and storytelling. Since Deal and Kennedy's pioneering work, the concept of organizational culture has attracted considerable attention, not least because of the breakdown of vertically integrated business structures and the emergence of trends towards increasingly dispersed and networked organizations. This is serving to dilute organizational culture and make it harder for managers to create a sense of corporate identity and cohesion. Today, it is highly likely that organizations will comprise a series of sub-cultures which will need to exist within a wider group culture. To help in managing this cultural kaleidoscope, it is useful to understand some of the main classifications of corporate culture which have been produced (see Table 6.2).

Organizational culture in practice

While the generic frameworks in Table 6.2 are interesting, more revealing is research into the cultural characteristics of innovative firms (Parnes et al. 1977;

Table 6.2 Organizational culture taxonomies

Deal and Kennedy (1982)	
Tough guy macho culture	High risk taking, high pressure, short term, quick returns, individualistic, success is winning and financial, heroes are survivors who win high stakes
Work-hard play-hard culture	Action orientated, high respect, low risk, success through persistence, lacking in thoughtfulness, customer focused, heroes are friendly
Be your company culture	High risk, low pressure, ponderous, deliberate, faith in good ideas, slow decisions, heroes are those who respect authority and show loyalty
Process culture	Classic bureaucracy, ritualistic, jargon bound, heroes are orderly punctual and attend to detail
Handy (1985)	
Role cultures	Classic bureaucracy, rule driven, hierarchical etc.
Power cultures	Based around one strong leader where status and control and obedience are highly valued, centralized decision making
Task cultures	Stressing flexibility, adaptability and egalitarianism, lateral communications and place high value on individual achievement
Person cultures	Highly decentralized and informal, control exercised through mutual agreement and partnerships and consensus
Graves (1986)	
Barbarian	Anti bureaucratic, workers are mavericks, workaholics, pop star individualists and contemptuous of a settled life, there is a constant atmosphere of ferment
Monarchial	Contempt for formality and planning, loyalty is valued, promotion from within, heavily depend on leader
Presidential	Democracy, an elected leader who embodies the wishes of the firm is sustained by subordinates but who needs to give clear messages to prevent people dissolving into sub groups
Pharaonic	Passion for order and stability, the system and procedures drive behaviour

Ekvall 1999; Sutton 2001; Robbins et al. 2003). This research shows that the cultures of innovative firms tend to be characterized by the following common attributes:

- Meaning
- Activism
- Listening
- Customer focus
- Experimentation and risk taking
- Trust
- Structured thinking
- Cellular structures
- Fluidity
- Divergent thinking and geekiness
- Creative conflict
- Collaboration and integration
- Challenging
- Fun
- Thrive on uncertainty
- Risk management
- Open-minded
- Flexibility
- Patience
- Empowerment
- Forward looking
- Tolerant of dissent
- Back themselves
- Resourcing
- Human resource development
- Creative spaces
- Broad unconventional metrics
- A buzz.

Meaning

Innovative firms see their business as a cause. This is not necessarily a social cause but anything that can make a special difference to their stakeholders' lives. They see their business as something important that is worth thinking hard about and building. This gives people a deep sense of responsibility to justify courageous revolutionary type actions. A common sense of identity around the cause means that people passionately connect with and believe in what they are doing. This provides the deep motivation that is required to push through the inevitable resistance of moving an idea to an end product.

Activism

Innovative firms allow and expect activism (the passionate promotion, championing and defence of ideas). They create a climate where positive debate around ideas is allowed and expected, where challenges can be voiced and where people can argue their case. There is respect for the individual and a competitive environment which pushes performance to new heights. Supported by an environment of positive debate, innovations are depersonalized with debates being focused on an idea not the person presenting it. Debates also tend to cut across regional, functional and chronological boundaries, and thinking about an idea can engage an entire company.

Listening

Creative firms 'listen' to outsiders who bring new ideas to their business from other contexts and industries. They are also open to the views of underrepresented voices such as young staff, newcomers and marginalized groups such as migrants. Innovative firms create a free and open market for ideas and talent with no prejudice about who is capable of developing a new winning idea and being supported. All that matters is the quality of the idea not who generated it. A good idea gets support no matter where it comes from.

Customer focus

Innovative firms tend to be highly customer focused. But they also recognize that customers cannot always imagine what is possible. So innovative firms do not always do what the customer wants – they seek to exceed expectations. They also seek to educate customers about the value of innovation and engage them in the process of creating and assessing new ideas.

Experimentation and risk taking

Innovative firms encourage people to experiment and take risks in an environment which is tolerant of failure and where mistakes are treated as learning opportunities not failures. Failure is recognized as critical to the creative process and it is assumed that many bad ideas must be generated to produce a good one. Even bad ideas can illuminate a problem and serve as a creative trigger. There is not an expectation that things will work first time but there is an expectation of learning through careful analysis. Innovative companies often extend this culture out into the marketplace by undertaking trials with customers and releasing many different products to see what sells and what does not.

Trust

Trust is the bedrock of creative firms and it works in two directions. First, people with ideas must feel their ideas will be respected and that they can speak their minds without fear of criticism or punishment. Second, firms must also trust that the person with the idea has the company's best interests at heart.

Structured thinking

Creative firms realize that creative thinking doesn't happen by accident and that it needs to be encouraged through deliberately structured processes. For example, Walt Disney has set-up three storyboarding rooms for his employees: the Dreamer Room, where all ideas can flourish and no criticizing is allowed; the Realist Room where those ideas can be given a reality test with no criticism; and the Critical Room where Walt Disney himself would constructively criticize what had been created. Ideas rotate through these rooms until a finished product is produced.

Cellular structures

Innovative firms tend to have a cellular structure. Innovations are spun-off into revolutionary cells which tend to emerge around new ideas as they develop and are proven. This structure give a company a small feel, even if it is very large. It frees people from a single business model and provides opportunities for people to nurture entrepreneurial talent where they are more able control the development of their ideas and benefit from them. It also ensures closeness to customers and maintains agility and focus as a company grows.

Fluidity

Innovative firms shift and evolve their structures, roles and responsibilities in response to evolving opportunities and to stimulate new approaches to engaging everyday problems.

Divergent thinking and geekiness

Creative firms employ divergent thinkers who 'care' about originality. Creative people, free radicals and misfits are not ridiculed but are seen as important to challenge ideas and provide fresh perspectives. While innovative firms also recognize that creativity needs to be eventually balanced by practicality, the constraints of implementation are not permitted to restrain early idea development.

Creative conflict

Creative firms strategically combine different roles, capabilities and personalities to induce creative tensions in teams. They recognize that team compositions that might lead to the most efficient ideas do not necessarily lead to the best ideas.

Collaboration and integration

Few firms have all the knowledge to innovate these days. So creative firms rarely work alone. Instead, they 'co-create' new ideas in collaboration with 'deeply' integrated supply and demand chains of complementary knowledge and skills. True 'deep' integration requires the sharing of risk and reward and an open culture which is driven by collaboration and trust rather than competition and confrontation. In innovative firms, knowledge is not owned by any one group but widely shared between all organizational levels and function (vertically and horizontally) and throughout the supply chain. There are minimal barriers to knowledge transfer and knowledge sharing and discourse helps facilitate learning and build consensus and ownership in shaping new directions.

Challenging

In innovative firms people feel challenged and there are opportunities for personal development and pursuit of passions which are not always directly and obviously related to work. People are allowed to grow personally and there is a basic drive to extend their personal boundaries, develop latent talents and explore new possibilities.

Fun

Innovative firms are often characterized by a certain child-like, but not childish, playfulness. People are permitted to 'try things out' without knowing what will happen and there is a fun-loving culture where humour is seen as productive and work is seen as fun.

Thrive on uncertainty

Innovative firms thrive on uncertainty and turn it to their advantage. They constantly seek to destabilize their markets through avenues such as regulatory change and thrive in risky environments.

Risk management

Innovative firms tend to be highly effective risk managers. They take calculated rather than blind risks and only move forward with a realistic and rigorous understanding of their exposure. Their higher ability to manage risk provides them with the confidence to be innovative and allows them to turn high-risk situations into opportunities.

Open-minded

Innovative firms take seemingly impractical, ridiculous and foolish ideas seriously. They recognize the value in applying things learnt in one realm to another completely different context.

Flexibility

Innovative firms set clear targets and objectives but provide people with autonomy to achieve them. The focus is on ends not means. Corporate objectives are designed to motivate, challenge and excite people rather than make them feel constrained and directed.

Patience

Innovative firms also recognize that discovering and developing ideas takes time and involves incubation. People are not tightly constrained to a full nine-to-five (or more) job. Firms trust employees and allow flexibility to enable people to go beyond basic ideas which don't always come within the 9 to 5 working day.

Empowerment

In innovative firms people are empowered to act in ways that are not tightly constrained by traditional competencies, roles and management oversight. People tend to have elastic job descriptions which provide them with the ability to shift their activities in response to opportunities and they are provided with the responsibility, authority and resources to effectively address the challenges they have been set.

Forward looking

Innovative firms are constantly scanning the business environment, seeking to understand and respond to it in advance. They look over the horizon rather than look at the horizon.

Tolerant of dissent

Innovative firms tolerate dissent. They welcome criticism and bad news. Harmony is not assumed to be associated with high performance.

Back themselves

Innovative firms do not just encourage new ideas but they also have the confidence to back them up. People at whatever level are able to get access to venture capital and resources to develop their ideas after they have been reality tested. Moral support is also provided because getting an idea from first notion to final product can be a long and arduous process with failures along the way.

Human resource development

Innovative firms continuously develop their people so that they are equipped with the knowledge and skills to innovate. Innovative firms balance technical competency development with skills in creative problem-solving. This provides employees with the ability to recognize problems and challenges, to find facts and information needed to resolve them and to identify solutions and overcome resistance to their implementation.

Creative spaces

Innovative firms generally exist in creative spaces which reflect their corporate culture and which help people (employees, customers and other stakeholders) identify with that culture. These spaces tend to be highly textural, engaging and fun. They help to break down organizational barriers by providing a flexible non-hierarchical environment where people can casually meet and they provide different types of workspaces where people can engage in varied types of thinking and workplace activities.

Broad unconventional metrics

Innovative firms tend to use a broader range of unconventional metrics to measure their performance. Success is rarely measured in purely financial terms and because they seek to differentiate themselves, innovative firms tend to develop their own distinct measures of success rather than those which their competitors are also using.

A buzz

Innovative firms tend to have an almost palpable buzz about them. This originates from the leaders of the organization who have a vision and sense of energy and enthusiasm which 'infects' others with a passion for creativity, success and learning.

Culture building strategies

In this section we explore a variety of strategies which have been successfully employed by innovative firms within and outside the construction sector to build an innovative culture.

Knowledge sharing

Many leading organizations have strategies, increasingly using social media, to enable staff to access and share information across a range of important business areas. For example, companies like Colonial First State, Allianz, Deloitte and ANZ provide regular social networking opportunities for staff through employee forums and clubs; peer-groups/networks, 'meet the CEO' events; and regular company functions to promote interaction and cohesion. IBM considers itself as part of an 'innovation ecology' which extends far beyond its business to business partners, universities, research institutes and clients etc. To facilitate knowledge exchange internally and externally across this diverse group, IBM uses an innovation knowledge-sharing portal to collect and disseminate its employees' and partners' best ideas. This innovation portal called 'Thinkplace' is open for only three days twice a year and receives tens of thousands of new ideas which are then discussed online, rated and scored and ranked by IBM's Emerging Business Opportunities (EBO) unit. Innovations generated through this process often become the growth areas of IBM and over the last two years, IBM's allocation of $70m in venture capital seed funding generated about $300m in new revenue. Anderson Consulting has a similar Learning Network called 'Knowledge X Change', Allen and Hamilton have 'Knowledge On Line', PWC have 'Knowledge View' and Hewlett-Packard has its own dedicated TV station that transmits to more than sixty business units worldwide. In the construction sector, Arup employs a similar strategy to IBM and has found that it is the new ideas generated by such forums that generate much of its business revenue. Arup staff are encouraged to develop their own professional networks with governments, universities and business partners in order to build up their personal knowledge and capabilities inside and outside the firm. They are also expected to share these contacts in a company-wide social networking site and in forums like the 'virtual global staffroom' designed to enable people from around the world to share ideas informally over lunch. What is evident, however, is that having a network is not enough in itself to guarantee success. The culture of the company must be open enough to support the network and it must be evident to employees and partners who are required to engage with it that the consultation process is meaningful and that ideas are seriously considered and some taken through to implementation with dedicated resources.

Skunks and bootleggers

Skunks and bootleggers are individuals or small groups who work on informal and often personal projects at the edge (or even outside) an organization's official functions. These side projects driven by passionate individuals are often very important sources of creativity and serendipity. The word 'skunkworks' originated in the seedy buildings or unused, quiet corners of offices where these people tend to work. And the term bootlegging comes from the tendency of these people to pursue these ideas in their own time and outside normal working hours.

Over the years, bootleggers have been responsible for some of the world's greatest advances in science and technology. For example, Air Commodore Sir Frank Whittle is credited with independently inventing the turbojet engine without official Air Ministry support. He and two retired RAF servicemen built his engine in their spare time and despite limited funding, a prototype was created, which first ran in 1937. Official interest was forthcoming following their success. However, the stress of creating the engine against all odds seriously affected Whittle's health, eventually resulting in a nervous breakdown in 1940.

Innovative companies such as 3M, Shell and Google also use bootlegging as a way to produce innovative ideas and to branch their business out into new growth areas. They allow staff to spend time working on their own promising projects outside their normal jobs. In Shell for example, projects considered to have potential can extend to one or two days a week and demonstration projects are normally used to test and refine ideas after they have been taken through several systematic evaluation stages. Other companies that allow bootlegging include Lockhead which regularly creates small tightly knit groups to work on special projects which are proposed by staff around personal interests.

Futures groups and think tanks

Many companies use think tanks as a forum for their most talented people to debate 'hot topics' and to think about over-the-horizon trends that may impact the business in the future. Such groups invite and listen to thought leaders, influence and lobby politicians and might also publish their deliberations, commission research and present them at conferences or government enquiries. For example, Arup has a 'Futures Unit' which is populated by its 'Arup Fellows' and co-opted thought leaders and futurists from outside the business. Laing O'Rourke has an Engineering Excellence Group which operates out of an Innovation Laboratory based in Sydney which serves its global business. The purpose of the Innovation Laboratory is to help the business win work and problem-solve. Financed by a distinct pool of funds which is ring-fenced to some extent from the rest of the business, the laboratory supports the company's strategic vision and serves as a resource for bid teams to help them differentiate themselves in the market.

Flexible job descriptions

To encourage innovation some organizations provide freedom for staff to explore new ideas and passions which may emerge into new areas of business. For example, Google has a 70/20/10 rule which requires all staff to work 70 per cent of their time on their particular job, 20 per cent on a project related to their job and 10 per cent on anything they want. This has led to many new products such as Google Earth (which came out of the 10 per cent allocation).

In companies like Google, staff can even create their own job subtitles to reflect the way they use their discretionary 30 per cent. It is this flexibility and the ability for people to define their jobs that contributes to the very high retention rates of these companies.

Free radicals

Some of the world's most innovative organizations have created 'free radical' positions for staff who have demonstrated a sustained record of invention and innovation which is recognized among their peers. These people are given considerable freedom to read, talk, wander and explore new business ideas and trends. In IBM these people are called IBM Fellows and are typically appointed for a five-year period with the objective to: shake up the system; be dreamers; heretics; mavericks; and geniuses. Appointment to one of these positions is recognition of high achievement and respect not only within IBM but in the wider community and has led to a number of Nobel Prizes.

Scholarships

Many organizations offer targeted scholarships to enable staff to take time out of the day-to-day pressures of work to pursue specific projects of interest to them and their business. Often, these scholarships are part of a talent management programme or are targeted at minority groups to support equity and diversity policies. For example, Rio Tinto and AURORA offer two indigenous scholarships to study at Oxford University. In construction, Brookfield Multiplex offer the annual John Roberts Scholarship to an outstanding candidate to study at an elite university of their choice overseas and to undertake research which is of strategic interest to the business.

Creative work environments

It has long been known that the physical work environment is crucial to a productive, healthy, happy and creative workforce. Many office buildings reflect the traditional rigid and planned approaches to strategy which we criticized in chapters one and two. They reinforce organizational silos and power structures by separating people into different spaces and levels and into functionally differentiated and easily monitored spaces which on the surface would appear to promote efficiency and productivity.

However, research into the relationship between buildings and organizational performance has transformed the way that businesses use space in recent years. Internal design now more closely reflects the principles of collaboration, identity and open communication which we have argued throughout this book are so essential to the innovation process. By providing open plan spaces where people can bump into each other (rather than formally meet), and by creating places which draw out people's creative capabilities, firms are able to encourage the mobility, diversity and insights which are the essence of serendipitous innovation. Frank Duffy's work has been especially influential in this area. For example, Duffy and Powell (1998) found that there are four types of space needed for people to function effectively and that this space can be provided physically or virtually:

1. Cells (where someone can be alone and in peace and quiet to reflect)
2. Hives (where people can work with 'everyone' in an open collaborative environment)
3. Dens (somewhere to take your team away to brainstorm)
4. Clubs (where people can meet voluntarily outside the structure of work to follow their interests with other like-minded people from anywhere in the business/supply chain etc.).

New approaches to using space such as activity-based working reflect these ideas of how to use space to encourage business innovation and creativity. For example, the new headquarters of Banking & Financial Service division of the Macquarie Group at 1 Shelley Street in Sydney accommodates 3,300 people in an Activity Based Working environment. Built by Brookfield Multiplex and designed by Fitzpatrick and Partners and Veldhoen and Company, the building was designed to empower Macquarie Group's employees by creating a learning environment which would help to change the business culture from a command and control model to a team-based model founded on collaboration and trust. The new Activity Based Work environment provides no occupant with an assigned desk. Instead, employees move through a variety of tailored work spaces as they perform different tasks throughout their working day. An employee has an anchor point, which is allocated as their 'home base' and it is here that their locker and storage resides. The design of this new building embraces new technologies such as laptops, follow-me printing, WiFi and IP softphones etc. This encourages increased collaboration and a mobile and flexible 'real time' work environment and post-occupancy evaluations suggests it offers significant improvements in perceived productivity and staff engagement.

Secondments and rotations

Secondment is the practice of one company making the services of an employee available to another company (or division) for a short period of time while continuing to remunerate the employee. Rotations on the other hand provide employees with short-term exposure to different parts of a business. Both secondments and rotations are a good way to break down knowledge silos, foster interdisciplinary relationships and build collaborative relationships across different business units or with key business partners which may lead to the development of new ideas. This enables employees to experience their company from another perspective, to foster collaborative relationships and to better understand the business challenges of those they have to work with.

Skills clubs

Skills clubs are effectively small informal networks between people with similar interests. The idea is that people from all organizational levels can share ideas and meet people of similar interests. Skills clubs are normally voluntary and staff can apply to create one. In Arup, for example, everyone is encouraged to be a member of at least one skills club. These clubs tend to focus on different key areas of the business and on key emerging issues. Over time, successful skills clubs become centres of expertise and develop ideas which drive new business opportunities. They also become a source of important knowledge to which others in the business can direct pertinent questions.

To facilitate an effective exchange of information, experience and expertise between these groups, many organizations have a searchable database of experts. This can be as simple as individual or group web pages which list people's skills, experience, presentations to external bodies, conferences, papers, interests and hobbies etc.

Study tours

Some companies such as Exxonmobil offer study tours of their own sites, of other sites and of companies considered as best practice etc. These can often be organized with business partners, providing an opportunity to forge stronger business and collaborative relationships.

Exclusive partnerships

Many organizations advertise 'exclusive partnerships' with well-known and highly respected organizations to attract creative people. For example, Australia Post advertises an exclusive partnership with Circus OZ to run its 'high flying leaders' leadership programme. In construction Brookfield Multiplex offers staff an opportunity to engage with issues of homelessness and long-term unemployment through its strategic relationship with the Salvation Army.

Libraries and resource centres

A library of information and learning material is essential to support learning and promote an innovative business culture. For example, DuPont have a library of material where employees can find information. This resource is also open to business partners and clients (to which they can also contribute), creating a perception of a learning partnership and of entering a learning environment when one engages with the company.

Journals and conferences

In order to encourage a learning culture and share knowledge internally and externally, a number of leading firms have developed their own internal peer-reviewed journal. For example, Arup has a journal called *The Arup Journal*. This is an internally peer-reviewed journal which allows staff with bright ideas to write about them. It also allows Arup to write up case studies on their projects and communicate lessons learnt etc.

Many companies also hold an annual conference (and sponsor external conferences) where staff and business partners are encouraged to submit papers to be peer-reviewed by an editorial board of external people and senior staff. These firms also invite prominent speakers from around the world to talk on a 'hot-topic' which is related to the business strategy.

Employing and using creative people

The most obvious way to increase an organization's creative abilities is simply to employ more creative people or use creative employees more effectively. However, it is not easy to identify creative people. For example, since creative acts demand special mental abilities we might expect that creative people have especially high IQs. Nevertheless, while psychologists have found that low IQ people do tend to be less creative than high IQ people, the relationship is very weak (Szilagyi and Wallace 1987). Furthermore, Belbin (1997) found that groups of 'clever people' generally performed badly, being characterized by destructive debate, intolerance and a lack of coherence. It is also interesting to note that many of history's most creative people, such as Copernicus, Rembrandt and Faraday, are estimated to have had IQs of less than 110.

Instead of depending on IQ, there seems to be general consensus that creative acts rely on many mental processes working together in harmony, namely: problem finding, idea generation, imagination, simplification, risk taking and motivation to learn (Santrock 1998). That is, creative people excel at finding problems, at finding new perspectives in their solution and at producing order out of chaos. They are also willing to take risks, to learn from failure and are determined, unconventional, self-confident, tolerant of ambiguity and intrinsically rather than extrinsically motivated by things such as money, evaluations, prizes etc. For example, one of Albert Einstein's main motivations in life was to find simplicity and to disseminate his ideas without undue publicity. Indeed, when asked why he used hand soap for shaving instead of shaving cream, he replied that using one bar of soap was less complicated. It is worth noting that a number of psychological tests have been developed to identify creative individuals, some based upon personality measures, some on biographical experiences, some on intellect and others on cognitive skills such as divergent thinking. However, there is controversy surrounding these tests and only inconsistent evidence of their ability to predict real world creative achievement (Wallach 1985). Consequently, contemporary creativity tests focus upon people's *outputs* rather than upon their mental abilities. For example, in rating Frank Lloyd Wright as one of the most creative architects of his day, one would tend to cite as evidence his buildings rather than his personality traits, although admittedly, he often behaved and wrote eccentrically.

Unfortunately, creative individuals like Einstein and Frank Lloyd Wright are rare and most managers need to elicit creative potential from specific combinations of relatively uncreative individuals. Belbin's (1997) research has been most influential in determining which combinations of roles, capabilities and personalities induce creative tensions in teams. Nevertheless, it remains the case that most management teams are created in haste without proper regard to such issues.

Creativity training

An alternative to recruiting creative people is to train an existing workforce to be more creative. Most creativity training programs are based upon the creative problem-solving (CPS) programme developed by Sidney Parnes (Parnes et al. 1977). This programme teaches participants a range of techniques to help them find facts, find problems, find ideas, find solutions and overcome resistance to their implementation. Unfortunately, while creativity training does seem to produce changes in creative ability, the effect is often short-lived (Basadur et al. 1986). Therefore, until there have been more long-term studies of its effectiveness, its value to managers remains uncertain and it needs reinforcing with other approaches to learning which can open people's minds to potential risks and opportunities.

One learning approach that could facilitate greater risk awareness in projects is *project closure reviews*. This simply involves spending some time at the end of a project reflecting upon the lessons learned and transferring them to people in future projects. For example, Ernst and Young use this technique by constructing a risk database using a standard set of questionnaires that are completed at the end of every project by different stakeholders. Future project managers can access this database to predict possible risks and develop effective management strategies, based on past company experience. In effect, this is a process of knowledge management. Similarly, British Petroleum have created a post-project unit with the sole mission of helping the company learn from its mistakes and successes. The unit's objective is to improve company performance and help managers more accurately formulate investment decisions, appraise them more objectively and execute them more effectively.

Academies

Many leading companies have invested large amounts of resources in the development of internal learning academies. For example, IBM's Academy of Technology was founded in 1989 and was modelled on the US Academy of Science and Engineering. Its purpose is to advise IBM's executives on technical trends and over-the-horizon risks and opportunities and to connect IBM technical communities across the world. Hewlett-Packard also has a learning academy with over eighty people directly employed to provide education for its own staff and those of its key business partners. This academy connects people from all parts and levels of the business and is not exclusive to managers. Shell Oil has also recently created a global Project Management Academy with the specific objective of supporting its capital projects programme and attracting, developing and retaining the industry's best project practitioners. A partnership with four university partners in Australia, UK, Netherlands and the USA, the concept of the academy is to provide an integrated development programme for Shell employees at all stages of the project life-cycle. The program includes learning events, assessment, accreditation, coaching and mentoring services, a global online learning network, community events and work experience opportunities. The university partners complement Shell's internal expertise with leading-edge research, education and training which is linked to Shell's strategic objectives. Shell adopts an open-book approach with their partners to enable them to fully understand Shell's business, its strengths and weaknesses and future challenges. In construction, Brookfield Multiplex has also recently established its BM Academy to integrate learning and development across its global workforce from graduate up to senior executive level.

Business schools

Another model used by many organizations is to establish a business school. For example, in Australia, Mirvac sponsored the Bond University Mirvac School of Sustainability, launched in June 2006 within the Faculty of Business, Technology & Sustainable Development at Bond University. This was a cross-faculty initiative which focussed on social and economic sustainability and sought to integrate environmental, urban planning and architectural disciplines. Another example is the Mace Business School in the UK which has been created to transfer knowledge and expertise to the Mace supply chain through the delivery of tailored programmes in management and construction. Mace argues that the Business School is a manifestation of the culture of learning that is promoted across all Mace projects and brings benefits to Mace customers by fostering integration and consistency in the approach to key management tasks and functions.

Corporate university

Typically, corporate universities follow two models: completely independent entities or; a formal industry-higher education partnership, based at a leading university which leverages off its facilities, existing courses, staff, library, quality control systems, education policies and strategies, accreditation agreements and academic networks etc. Growing in popularity, particularly in the US, the main advantage of corporate universities is that they use the resources of traditional universities but can tailor their courses to the strategic needs of the business. Firms which have created their own corporate university include: Land Rover, AT&T, Amoco, NCR, Rover, Xerox, Saturn Corporation, McDonalds, Arup, Motorola, Disney, Rover, Xerox, Andersons, American Airlines and Bell Atlanta.

High involvement workplaces

The concept of high involvement workplaces (sometimes called high performance or high commitment workplaces) has attracted considerable attention in recent years, building on the simple insight that people are more likely to give of their best if they feel valued and are given the opportunity to contribute their ideas to secure their employment in a business (UKWON 2007). Developed initially in the 1980s in response to the perceived success of Japanese management practices, policy-makers around the world have also embraced the idea as a means of achieving high employee engagement. As MacLeod and Clarke's (2010: 8) report to the UK Secretary of State for Business concluded,

> engagement is about creating opportunities for employees to connect with their colleagues, managers and wider organization. It is about creating an environment where employees are motivated to want to connect with their work and really care about doing a good job.

Not only does higher engagement result in higher productivity, safety, staff retention and staff advocacy of their employer but it also correlates strongly with innovation. Macleod and Clarke's research showed that 59 per cent of engaged employees said their work brought out their creativity compared to 3 per cent of disengaged employees. They cite many examples of highly innovative organizations that focus on engagement including Microsoft and the Mace Group, an international consultancy and construction company with £700m turnover in 2009 and 2,800 staff in 35 countries. Mace has developed its own engagement model which aims to attract and retain the best people. The model focuses on providing employees with interesting work, corporate social responsibility, two-way communication and training and development. Employees are encouraged to rotate around the company to vary their work and support their development and are given responsibility for decision-making and the flexibility to deliver in ways they see fit. Mace has also developed leadership programmes and skills training for its staff and has developed a managers' charter which requires line managers to take responsibility for the development of their subordinates. In general terms (although it is still open to debate), the high involvement workplace approach challenges traditional functionally siloed, top-down, command and control type structures and management practices. Instead it promotes high levels of involvement, participation, adaptability, flexibility and empowerment across all levels of organization. These workplaces are characterized by a focus on the psychological contract rather than the formal employment contract in order to promote positive workplace relations. This is done through treating employees as entrepreneurs and by devolving decision-making power to individuals and groups closer to the customer interface through flatter networked forms of organization. This in turn requires that staff are better developed and given the skills and knowledge to make effective decisions. In essence, high-involvement workplaces are designed to be knowledge-based organizations which use management practices like team-working, staff briefings and problem-solving groups.

The rather utopian concept of high involvement workplace is, in many ways, a rebadging of Burns and Stalker's (1961) 'organic' organization. Like Burns and Stalker, proponents argue that it offers mutual gain for business and employees through greater levels of worker empowerment, participation, commitment and staff development. This leads to higher levels of productivity, innovation and profit. It is also argued that this approach creates a work environment which better meets the expectations of Gen Y who do not want to be treated as cogs in a machine but who want to have a say in how their company is run.

Although there is some evidence to support such assertions, it is important to remember the point made at the start of this chapter – that there is 'no one best way to manage'. There is little doubt that this way of doing business would be less effective in some contexts than others and may even be counterproductive. Furthermore, while advocates claim that the links to positive performance outcomes are unequivocal, skeptics argue that much of the data collected has been crude and largely focused on the manufacturing sector. Its transferability to industries like construction would therefore have to be investigated before it is adopted. For example, while high involvement organizations tend to be associated with higher levels of training for their employees, it is not clear how this translates into higher skills levels and then into firm performance. Furthermore, there is some evidence to suggest that to prevent poaching by competitors once employees have been trained, firms tend to invest only in non-transferable skills development and leave transferable skills development to employees. Finally, there is some evidence to suggest that high involvement workplaces are associated with lay-offs and have no 'tangible' benefits for employees other than more intrinsic rewards such as greater job satisfaction and autonomy etc. In fact, some research has shown that high involvement workplaces lead to lower wages and can create higher levels of stress, work intensification and insecurity for some employees associated with higher levels of personal accountability that accompany more democracy and empowerment and the lower levels of management oversight and control.

National culture and innovation

If organizational culture in central to innovation does this translate to national culture? In other words, are some countries more innovative than others and how does this affect firms which are incorporated there? Certainly, as we showed in chapter two, some countries are ranked more highly than others on various national innovation indexes such as the Global Innovation Index. However, only one index, the Economist Intelligence Unit's Innovation index, considers the national culture as an important variable in its national innovation rankings.

Despite absence of national culture as a factor in international innovation rankings, there has been a considerable amount of anthropological research to indicate that national culture might be an important factor in a firm's capacity to innovate. To anthropologists, national culture is the common set of beliefs, behaviours, objects, symbols, language, customs, values, norms and rules which bind people together in a country. These are learnt and instilled through the process of socialization in institutions such as families, schools, religions, governments and companies. While the concept of national culture has been somewhat eroded in recent decades by globalization, research shows that people still recognize their national culture as an important source of identity. Furthermore, it remains an important aspect of organizational life as firms increasingly expand their operations into overseas markets and as workforces become more culturally diverse in response to growing global labour mobility.

Perhaps the most well-known classification of national cultures is the one produced by Geert Hofstede (1980). Hofstede founded and managed the Personnel Research Department at IBM during the late 1960s and played an active role in the introduction and application of employee opinion surveys in over seventy national subsidiaries around the world. Travelling extensively across Europe and the Middle East he interviewed many people and conducted surveys with over 100,000 employees, classifying forty different national and organizational cultures across four dimensions of culture:

1. Power distance (degree to which power is accepted)
2. Masculinity/femininity (degree to which dominant values are masculine/materialistic or feminine/caring)
3. Uncertainty avoidance (degree to which people feel threatened by uncertainty)
4. Individualism/collectivism (degree to which people value the group).

Given the importance of dissent, risk taking and collaboration to the innovation process, Hofstede's work clearly indicates that some national cultures are more likely to be conducive to innovation than others. This is confirmed by a recent review of innovation in China by Orr and Roth (2012) who found that one of the biggest obstacles to innovation in China is Chinese workers not being as comfortable with risk taking and failure as Western companies. Orr and Roth found that a culture of obedience and adherence to rules prevails in most Chinese companies where breaking or even bending the rules is rarely tolerated. The Chinese aversion to 'loss of face' also means that failure is scorned upon in many companies and that employees are discouraged from experimenting and taking initiative. Orr and Roth also found that while Chinese firms have become highly adept at delivering products in large volumes and at the rapid commercialization of new products, their ability to foster cross-company and supply chain collaboration has not kept pace. While we still have only a rudimentary understanding of how differences between national cultures might affect levels of innovation in organizations, this research is important in alerting managers to the fact that their management strategies are often based in Western cultural connects and might not work as effectively in other national contexts.

Learning from innovative people

In the same way as we can learn about innovation from the cultures and practices of creative organizations, Dodgson and Gann (2010) show that much can be gleaned by studying the lives of innovative business people throughout history. For example, much can be learnt from industrialist Josiah Wedgwood (1700–1795) who made good design and high quality ceramics available for the first time to a mass market by bridging the artistic and scientific communities with a completely new approach to industrial design. Wedgwood collaborated with artists, sculptures, scientists, engineers, industrialists and businessmen and architects and was driven by a constant search for new ideas to improve the quality of his products. Wedgwood was passionate about the power of improvement from experimentation and was the first to use steam power to mechanize the process of throwing and turning pots, reducing labour costs, increasing efficiency and improving the quality, reliability and productivity of his business. As his business grew and demand exceeded capacity, he outsourced production to other potters and innovated in managing his supply chains by forging strategic partnerships with suppliers around the world to ensure the highest quality and reliability of raw materials. Employing Adam Smith's ideas around the division of labour, Wedgwood transformed a traditional cottage-based potting industry into a global manufacturing business where people could specialize to improve skills and quality in specific areas, rather than having to master all aspects of design and production. The efficiency of Wedgwood's factories was driven by his obsession with costs. He studied cost structures to ensure economies of scale, avoided one-off designs and mass produced many of his products, keeping the basic product the same but embellishing them with motifs to reflect changing fashions. While Wedgwood was very cost conscious, he refused to compete on lowest price. He paid higher wages than his competitors, invested in training and skills development and was very conscious of the health, welfare and safety of his workers. Wedgwood believed strongly in the power of networks. Outside work he was a member of many learned societies and mixed frequently with innovative businessmen from other industries who were highly influential over his thinking. Wedgwood's marketing was also innovative for the time. He sought patronage from influential people like politicians and aristocrats, using their endorsements in his advertising. Using his aristocratic connections, he built a reputation for quality in foreign markets and displayed his products in detailed catalogues with money-back guarantees and free delivery, backed up by a travelling salesmen service. But while Wedgwood was intensely competitive, he was also prepared to share ideas. He disliked patents and encouraged collaborative research with competitors and partners to solve common problems across and within industries.

Chan and Cooper's (2011) analysis of what drives modern thought-leaders in the UK construction industry is not dissimilar to Dodgson and Gann's (2010) analysis. The common factors which link the leaders interviewed by Chan and Cooper are a deep thirst for new knowledge and a passion for seeking improvements in the way the industry works. People lie at the heart of what these leaders do and there is a deep belief in the value of integrating knowledge from many different sources. In their early lives, many of these leaders were mavericks and rebels who were prepared to question conformity, authority and the status quo. Of particular importance was the inspiration afforded from a sense of place and time, by travel and from formative events and experiences during their lives when they were inspired by remarkable people who gave them the passion for new ideas which followed them through their lives. Finally, in successfully implementing their ideas, these industry leaders were also able to use and mobilize influential networks to energize activism for their cause.

The growth of managerialism

Today, the world is a very different place than it was in Wedgwood's time. Wedgwood lived at the dawn of the industrial revolution when there were few constraints on managers in the way they treated their workers and organized their businesses. In contrast, modern managers have to operate within a far more regulated, educated and empowered society with many moral and legal expectations of behaviour. Kay (2013) provides a timely example of this in celebrating the 150th anniversary of the London underground sewage system. As he points out, in 1858 the Thames River was an open sewer and abnormal weather produced what became known as 'the great stink' which made much of London unbearable to live in. After widespread complaints that something had to be done, the next decade was spent building an extensive network of sewers under the supervision of engineer Joseph Bazalgette. The creation of the London Embankments also enabled the building of the London Underground and while time has ensured that London has benefited massively from the system (which has only just reached its capacity), Kay (2013) speculates that it probably wouldn't have been built if it had been subjected to modern Treasury and planning approval processes. The construction of the sewers had many negative consequences such as the destruction of ancient gardens and buildings and economists would have been required to calculate the return on investment from such a scheme which they would inevitably have got badly wrong. Economists could never have predicted the impact that the system had on the health of the population, on transportation and on the national economy. As Kay points out, the salient fact is that London could never have become a great business and financial capital if its early development had faced the multiload of regulations we face today (Kay 2013: 9).

While regulation has been necessary to protect society from the worst excesses and abuses of development and business activities, Locke and Spender (2011) argue that we now live in a world where accountability has gone mad and where innovation is smothered by the need to measure and monitor business activity in every detail. Locke and Spender argue this lies at the heart of the economic and social crisis the West faces today and has been driven by a pervasive fear of litigation or punishment for doing something wrong, by the dispersion of business ownership and associated short-term speculations on global financial markets and by increasing red tape and green bureaucracy.

Even in academia, which should be a bastion of open and free thinking, Braben (2004) argues that people are under threat from an insidious tide of laws, rules and regulations to which they must conform. These controls are administered through growing research bureaucracies which set national priorities to which researchers must conform and which link university performance to publications in incestuous academic journals which in many cases sterilize new ideas by forcing them to conform with badly managed peer-review processes. Braben argues that this system of 'quality control' force ideas to coalesce around a central norm and misunderstands the explorative nature of research. It also ensures that researchers who operate outside these artificial and narrowly defined frames of reference, which are normally informed by present rather than future needs, have great difficulties getting funding for their work. Inevitably, Braben argues that this system is destroying the trail of dissenting, idealistic and eccentric people such as Copernicus, Kepler, Galileo, Luther and Einstein who, throughout history, have been responsible for most of the advances in science and technology we see today.

The importance of dissent

Organizations are a site of continual conflict between the forces of conformance and dissent. While conforming behaviour is essential to the smooth running of any business, individual acts of dissent are also essential to drive the innovation process forward. As Jonathan Ive, Vice President of product design at Apple Computers, once said 'Innovation takes an unreasonable man'.

According to Hornstein (1987) there are many subtle and invisible forces that stifle dissent in organizations. As Sage and Dainty (2012) found in their study of power in one architecture practice, the imposition of control can often be hidden, subtle and symbolic despite the rhetoric of company statements which often suggest open business cultures that have little formal hierarchy or power structure. However, most organizations encourage conformity through 'formal' rules and procedures, by punishing those who step outside them and by rewarding those who don't. Consequently, most people conform to the status quo and current consensus despite the fact that almost everyone knows that all is not well. Ryan and Oestreich (1998) found that the most dangerous side effect of such an environment is a fear to 'speak up' and of working where relationships become clouded in 'undiscussables' (organizational secrets that everyone knows about but no one talks about openly). According to Ryan and Oestreich the most common undiscussables in organizations are, in order of priority:

1. Management performance
2. Coworker performance
3. Compensation and benefits
4. Equal employment opportunities

5. Change
6. Personnel systems other than pay
7. Individual feelings
8. Performance feedback to respondent
9. Bad news
10. Conflicts
11. Personal problems
12. Suggestions for improvement.

Using dissent as a classification system for management styles, Hornstein found that there are typically four types of managers in organizations:

1. Courageous managers
2. Confronting managers
3. Disaffected managers
4. Conforming managers.

Courageous managers are in the minority (about 20 per cent of the management population). They believe that organizational regeneration is desirable and possible and are prepared to do something positive about it. For these managers self-identity, reputation and career is tied to the organization and its performance. They talk of 'we' or 'us' when they discuss the business and if they believe that something is wrong with it, then they cannot live with themselves if they do not speak up and do something about it. To them, silence is equal to endorsement and they feel compelled to speak out. And because they only do so after careful consideration of their position, their chances of bringing about change are generally very high.

Confronting managers also believe that change is needed and possible. However, they seek to blame their bosses for the problems and move against parts of the organization they dislike. These types of managers tend to focus their courageous act on a particular issue or a particular superior and their actions tend to be angry, irrational and passionate. Furthermore, they tend to be aimed at unseating a superior enemy who they believe has harmed their organization. For this reason, they are the most unsuccessful of all managers.

Disaffected managers think that organizational regeneration is not possible and move to psychologically isolate themself from the source of the problem or resign and move to another organization. These managers are weighed down by feelings of helplessness and hopelessness. Having been punished and discouraged by the forces of continuity, these managers have surrendered to the system and stopped caring about the organization they work for.

Conforming managers support whatever policies and practices their organization is currently promoting and never question the status quo. They go with the tide, loathe confrontation, love approval and their vision of the future is a reflection of the organization's past.

Organizing for dissent

We have argued in previous chapters that innovation today should be an open, collaborative process driven internally by organic business structures and externally by strategic supply chain alliances. The theory is that by constructing a group of firms and people with complementary skills and knowledge, barriers are broken down and knowledge from diverse areas is combined in new ways to create teams where synergies are realized that are beyond the sum of the capabilities of the individuals acting alone (Steve and Thomas 2008).

While the theory sounds good, effective teams are not as easy to create as they may first seem. For example, Meredith Belbin, a British researcher and management theorist who has studied team-work for many years, famously observed that people in teams tend to assume different roles (Belbin 1997). He defined a team role as a tendency to behave, contribute and interrelate with others in a particular way and named nine roles that he argued underlie team success. He categorized those roles into three groups: Action Oriented, People Oriented and Thought Oriented. Teams can become unbalanced and ineffective if all team members have similar roles. If team members have similar weakness, the team as a whole may tend to have that weakness. If team members have similar team-work strengths, they may tend to compete (rather than co-operate) for the team tasks and responsibilities that best suit their natural styles. Belbin's team roles are described in Table 6.3.

Psychologists have also explored the dysfunctionalities which tend occur in homogeneous groups. For example, Professor Irving Janis, a research psychologist from Yale University, coined the term 'groupthink' to refer to the phenomenon which occurs when homogenous groups castigate dissenters and develop a close-mindedness which leads to a loss of individual creativity, uniqueness and independent thinking.

Janis documented eight symptoms of groupthink:

Table 6.3 Belbin's team roles

Role type	*Role sub-type*	*Role*
Action Oriented Roles	Shaper	Challenges the team to improve.
	Implementer	Puts ideas into action.
	Completer-Finisher	Ensures thorough, timely completion.
People Oriented Roles	Coordinator	Acts as a chairperson.
	Team Worker	Encourages cooperation.
	Resource Investigator	Explores outside opportunities.
Thought Oriented Roles	Plant	Presents new ideas and approaches.
	Monitor-Evaluator	Analyses the options.
	Specialist	Provides specialized skills.

(Adapted from Belbin 1997)

1. Illusion of invulnerability – excessive optimism that encourages extreme risk taking
2. Collective rationalization – supporting each other's views and opinions
3. Belief in inherent morality – belief in the rightness of their cause and ignorance of the ethical or moral consequences of decisions
4. Negative stereotypes of out-groups – outsiders are the enemy and are ignorant
5. Direct pressure on dissenters – punishment for those who express alternative points of view
6. Self-censorship – many unspoken truths
7. Illusion of unanimity – everyone assumes everyone else agrees
8. Self-appointed 'mind-guards' – leader is protected from bad news.

Other researchers have expanded our understanding of how homogenous groups encourage potentially dangerous conforming behaviour. For example, Paulus et al. (1993) have highlighted the process of 'evaluation apprehension' where individuals tend to censor their own ideas because they want to make a good impression on other more senior group members. And there is also a 'social loafing' effect where some people tend to leave it to others in the group to do the thinking.

These insights into the potentially destructive dynamics of group behaviour are not new and go back to Le Bon's (1908) famous study of events during the French Revolution. His work first revealed the potentially ugly side of groups, which he argued were always intellectually inferior to the individual. Groups, he showed, could cause people to behave in extreme and sometimes violent ways that wouldn't normally be acceptable to them or to society. This happens because the crowd affords people anonymity which allows them to hide behind others' behaviour. Freud (1921) later argued that 'the crowd' unlocks the unconscious mind and has a 'hypnotic' effect on individuals, reducing their super-ego (the sense of right and wrong), bringing the 'id' (people's basic natural instincts) to the surface and reducing their ego (the modification of our basic instincts by society). Freud argued that as group members lose their own identity and replace it with a group identity (a process called de-individuation), there is a lowered sense of individual responsibility for individual actions.

The invisible expectations and pressures which encourage people to conform to group expectations are referred to as a group's 'norms'. These norms can exist in groups, even many years after they have disbanded, although it has been found that their influence can be reduced dramatically by some simple management strategies:

1. Avoid homogeneous groups of similar knowledge, values and beliefs.
2. Leaders should resist from stating preferences and expectations at the outset of group tasks.
3. Clear task objectives should be set for the group (norms tend to develop in response to ambiguity).

4. Each member of the group should be encouraged to discuss the group's deliberations with outsiders and report back to the group on their reactions.
5. One or more experts should also be invited to each meeting on a regular basis to act as devil's advocate, to challenge members' views and highlight alternative courses of action.
6. Use reward systems to positively reinforce dissent.
7. Simply increasing the size of a potential minority by one member dramatically increases resistance to majority influence.

Encouraging non-conformist thinking

Since human beings are social animals, other people's opinions can have a profound effect on the way we think. Most people want to be seen as a team player, they want to identify with a cause or group and they do not want to be seen as the 'odd one out'. This conforming behaviour is particularly relevant to our understanding of the innovation process because innovation inevitably requires people to challenge accepted ways of thinking. As Greenstein (1998) found in his biographical study of great thinkers throughout history, people who think differently often have to fight powerful social, cultural and political forces. However, people who are prepared to stand up against the prevailing view are rare and most people tend to feel very uncomfortable with ideas which do not conform to their existing belief systems and tend to avoid them – a well-known and very powerful psychological effect known as 'cognitive dissonance' (Festinger 1957).

Aronson's (1999) research shows that in seeking social confirmation and reinforcement of our own views, we are most easily influenced by those we see as experts, those we like, those we consider trustworthy and those who we socially identify with (normally through group membership). Interestingly, Wilder (1977) found that pressure to conform is greatest in groups of three to seven members. In larger groups people become suspicious of unanimity. Research also tells us that relative 'power' is important to whether people conform or not and French and Raven (1959) identified five sources of power that a person can use to persuade others to adopt their point of view, namely:

1. Coercive power – derived from someone's ability to punish others for not conforming
2. Reward power – derived from someone's ability to reward for conforming behaviour
3. Expert power – derived from a person's recognized expertise and credibility as a source of information
4. Referent power – derived from a desire to be associated with that person or source
5. Legitimate power – derived from a person's official formal authority to instruct you to believe them.

These categories of power are useful in understanding how to stimulate more innovation in the construction industry. For example, clients are in a powerful position to drive innovation in the industry through their substantial reward power base. Governments can also use their coercive power to force change through legislation and regulations. And the UK has shown that innovation champions like Latham and Egan can have a significant influence on levels of innovation by using their legitimate power-bases to drive change. Finally, the referent power of benchmarking clubs has also been used around the world to create an important association of innovators with best practice. These strategies reduce the 'pariah effect' of innovation. As Aronson's (1999) research shows, innovators are often seen negatively by established groups as a deviant people or fringe dwellers, who need to be brought back into line. Innovators acting alone can rarely resist pressure to conform, but as part of a minority group (for example, a benchmarking club) they are more likely to be tolerated and may even have the opportunity to influence the majority's view of the future.

Creative thinking

We pointed out in chapter one that creativity involves generating new ideas while innovation involves putting them into practice. In seeking to understand the creative mind, psychologists have found that we are all born with two contradictory sets of cognitive instructions (Csikszentmihalyi 1996):

1. A conservative tendency (made up of instincts for self-preservation and saving energy) and
2. An expansive tendency (made up of instincts for exploring, creativity and for enjoying novelty and risk).

Creative thinking was long seen by Gestalt psychologists as something done only by highly intelligent people which involved an inspirational Eureka moment of insight. However research into the lives and work of highly creative people has debunked these two common myths. It is now widely accepted that creative thinking involves four distinct stages (Gardner 1993):

1. Preparation – exploring the problem-space through the creation of a conceptual model
2. Incubation – the problem is often left alone for a while and reflected upon through unconscious thinking by delving into long-term semantic memory
3. Illumination – when a solution or critical insight arises at an unexpected moment
4. Verification – the time-consuming and laborious process of testing, evaluating and refining an idea over and over again to ensure it is valid and reliable.

In addition to discovering that creativity is a structured rather than random process, the widely assumed link with IQ has also been debunked. In fact, research shows that IQ and creativity could be inversely related. This is because IQ measures 'convergent' thinking whereas creativity requires 'divergent' thinking. Convergent thinking involves solving well-defined problems in a linear and rational manner under the assumption that there is one correct answer 'out there' waiting to be discovered. In contrast, divergent thinking assumes that solutions are context-specific and involves; generating a great quantity of ideas; switching from one perspective to another; and constructing unusual associations.

So it would seem that most people are capable of being creative with the right environment and training. To this end, Imber (2011) proposes fifty 'creativity boosters' which can be used by people to enhance their creative capacity. Some of the simple creativity techniques which Imber proposes include:

- Five-minute 'brain warm-up' sessions before meetings.
- Avoiding giving employees open-ended tasks.
- Bringing in 'outsiders' to provide an alternative point-of-view.
- Providing general objectives rather than specific objectives to encourage people to think more broadly.
- Broadening people's horizons by exposing them to ideas that they wouldn't normally see.
- Creating a conducive office environment to stimulate interactions and lateral thought. This may involve the use of colourful décor, furniture, informal meeting places, variety of texture, artifacts such as toys which encourage creativity and pictures that emphasize the importance of being original in work.
- Giving people challenging tasks with a realistic stretch factor.
- Avoiding 'pay for performance' systems such as sales targets which can prevent people from engaging in innovative solutions that may temporarily threaten those rewards. Research has shown that creativity is enhanced by tipping the balance back towards intrinsic rewards such as recognition.
- Recruiting creative people.
- Training existing staff to be more creative.
- Structuring teams to encourage creativity. Teams of about six to nine members have been found to be optimal in most situations and there needs to be a mix of knowledge and personalities to create the constructive conflict and variety of insights which is the essence of the creative process.
- Rotating the membership of existing teams that may be tired or to take them out of their normal context to stimulate new thinking and relationships.
- Avoiding onerous time pressures – people need time to reflect when being creative.
- Relaxation is important to stimulate the creative mind and to this end, firms can provide lunchtime seminars, mandatory tea breaks or yoga classes to enable them to do so.
- Mentoring can also be useful, as can the more unusual technique of reverse

mentoring where senior employees are mentored by a younger employee who might be more informed of certain technologies and ideas which senior managers are often insulated from by their position.
- Making sure your employees are happy – one's emotional state is strongly linked to creativity.

Dreaming and imagination

Gary Hamel (2002: 11), one of the world's most influential management thinkers, has argued that that there has been 'a massive failure of collective imagination' in the business world. While reliance on past experience to forecast the future may have served us well in the past there is an increasing body of knowledge which indicates that our future poses us with completely new scenarios that past experience cannot accurately inform. One of the best illustrations of this limitation is the challenge of climate change, where we have virtually no precedent to work from. In other words, looking in the rear view mirror is no longer a reliable way to inform the future. We have to become better at using our imagination to try to visualize what the future might look like.

Using our imagination requires a type of thinking called 'deductive reasoning' which Hamel (2002) argues is rare in most organizations. In contrast to 'inductive reasoning' where one moves from observations and measures of past patterns and regularities in data to forming general conclusions or theories about the future, deductive reasoning works the opposite way, working from what might happen in the future back to the present. Scenario analysis is a common form of deductive thinking increasingly used in business to speculate and imagine what the future might hold. This involves creating alternative possible future scenarios ('alternative worlds') which might describe a pessimistic, optimistic and most likely picture. In contrast to the technique of forecasting which relies on the inductive logic that past observations will still be valid in the future, scenario analysis does not rely on extrapolating from historical data. Instead, it uses people's experience, judgement and creativity to imagine possible future scenarios which may only be connected tenuously to the past.

To help us understand the process of imagination further, it is useful to refer to the famous work of Sigmund Freud (1954), the Austrian neurologist who became known as the founding father of psychoanalysis. Freud famously identified two types of thinking:

1. Primary process thinking
2. Secondary process thinking.

Primary process thinking is instinctual (driven by the *id*) and manifests itself in dreams and daydreams. In contrast, secondary process thinking operates within the constraints of the real world, is goal directed and is used for everyday problem solving. Freud was the first to recognize the importance of dreaming to the creative process. And this is supported by the history of innovation which shows how many of history's greatest innovators have cited their dreams as an important source of imagination and inspiration. For example, Einstein was notorious for his wandering mind and it is widely thought that his theory of relativity was conceived while daydreaming. Bill Gates has also stated that he first imagines his products before they become reality and Australian born Nobel prize winner Elizabeth Blackburn told a recent audience that daydreaming has undeniably contributed to her success (MacGibbon 2010).

While today's culture of production and ruthless efficiency derides daydreaming, seeing it as a lazy habit which shows a lack of discipline and poor self-control, recent research in the field of cognitive psychology has portrayed the act of daydreaming in a far more positive light. While much of the evidence linking daydreaming and creativity remains anecdotal, researchers are beginning to find experimental proof of a significant positive relationship. For example, Cohen and Schooler (1996) have shown that people who engage in more daydreaming score higher on experimental measures of creativity. Daydreams appear to involve a more relaxed style of thinking which is uncluttered by the worries of everyday life and which enables people to better contemplate ideas that would otherwise seem silly or far-fetched. Common anecdote again supports this finding. For example, it is not unusual to hear someone say that an idea came to them when lying in the bath or standing in the shower. However, Cohen and Schooler also found that individuals who are unaware of their own daydreaming exhibit lower levels of increased creativity than those who are consciously daydreaming. So it appears that it is not enough to just daydream. Rather, people must maintain enough awareness so that when they start to daydream they can interrupt their thoughts and notice a creative insight when it happens.

Since this early pioneering research, it has become widely recognized that daydreaming is a fundamental feature of the human mind – so fundamental, in fact, that it is often referred to as our 'default' mode of thought (Mason et al. 2007). Many scientists now argue that daydreaming is a crucial tool for creativity, self-regulation, problem solving and thinking about the future. In contrast to the past, daydreaming is now seen as a sign of a healthy human brain. Daydreaming allows the brain time to make new associations and connections, to order thoughts and to subconsciously reflect on things and imagine new ideas that might not yet exist. So in contrast to traditional beliefs that our brain is doing nothing when daydreaming, it is really doing a tremendous amount of significant value.

In seeking to change the lack of imagination which afflicts industry, French philosopher Michael Foucault's writings on power, knowledge, and discourse are particularly useful. According to Foucault, if predictions about the future are to be taken seriously by society then the past must represent the foundation against which they are anchored. If they are not he argued, then the ideas will be accused of being 'foundationless'. In other words, people look to the past to sort new ideas, place them in context and to provide an 'evidence-base' from which to draw conclusions. This process of inductive reasoning which is the foundation of scientific method and thinking in almost every Western society has ensured that history represents the 'field of knowledge' with which we all identify. Foucault's work tells us that the popularity of new ideas in society is likely to be determined by the extent to which they match our current preconceived beliefs and needs. But it also shows that any new idea's acceptance will be determined by the power of vested interest groups which have a stake in protecting existing versions of the truth. Foucault argued that it is these interest groups that determine which knowledge 'rises to the top' of society to become an accepted part of our history. Conversely, it also determines which knowledge is suppressed. The forces of inertia which Foucault pointed to in broader society are also evident in the way that knowledge has been forged in the construction industry. For example, Harty et al. (2007) analysed a decade of 'future studies' published in the construction sector and found that they were more about the present than the future and have been largely unable to predict the transformative discontinuous 'wild card' or 'Black Swan' events that affected the industry. Chan and Cooper's (2011) more recent analysis of future studies in construction argue that this convergent thinking arises from the small number of elite industry stakeholders who repeatedly compile these reports whose value systems and power relations at the time undoubtedly influence whose views get represented. So it would seem that if we are to be better at predicting the future and imagining what it might hold, then we must abandon our attachment to evidence to verify the validity of new ideas and question the predictions of vested interest groups who may have an interest in seeing a particular future come to fruition.

Innovation requires new types of thinking

If traditional management thinking does not provide suitable tools to help us innovate in today's uncertain, dynamic and interconnected world, then what type of thinking will?

Design thinking

To answer this precise question, Le Masson et al. (2010) and Smith (2011) propose an alternative approach based around 'design thinking' which requires us to think about products and services as having 'unstable identities'. This is based on the observation that the identities of new products are changing ever-more rapidly and that the traditional boundaries between products and services are also changing through a constant process of creative destruction. For example, in recent years the modern phone has become a mobile device which incorporates a camera, a computer, a games station, a voice recorder, an alarm clock and something to pay the bills – eliminating traditional markets for products in these areas. Richmond (2012) points out that in 2012, sales of point-and-shoot cameras fell by 30 per cent as people opted instead to use their phone camera, leading to the recent demise of iconic firms like Kodak which failed to capitalize on the digital revolution. The automobile industry is going through a similar revolution with the development of hybrid engines and firms in this industry that do not respond to the shift will likely suffer a similar fate to Kodak.

The common attribute of these new integrated products and services is that they combine ideas and technologies from different industries in ways which have not been seen before. For example, in the construction sector, the embracement of sustainability is seeing the integration of new technologies from a wide range of industries which have traditionally had little involvement in the industry. So today, innovation is being driven by firms who challenge traditional product identities to create and capture new integrated market identities through hybrid products and services with multiple functions. Le Masson et al. (2010) claim that this hybridization of products and services is not a passing trend but a permanent and fundamental structural shift being driven by customer demands for more integrated, more simple and more efficient multifunctional products and services.

Intuitive thinking

In chapter one, we briefly discussed the importance of intuition in entrepreneurial decision-making. In the context of innovation, intuition refers to an entrepreneur's inherent ability to see value in unexpected connections and insights without clear evidence or reason. Like the concept of luck, an entrepreneur's intuition might seem magical, but as Daniel Kahneman (2011: 10) argues in his most recent international bestseller *Thinking Fast and Slow*, this is a myth. In untangling how intuition works, Kahneman, an American Nobel Prize winning psychologist, draws together recent developments in cognitive and social psychology to show how experts who possess intuitive capabilities have become so familiar with their particular business that they come to see it differently from the rest of us. The chess master can see moves that the layperson cannot, the master carpenter can sense that a piece of wood will splinter when an apprentice cannot and an entrepreneur can sense an opportunity where others cannot. For example, one of the world's most charismatic entrepreneurs Sir Richard Branson openly acknowledges that while he analyses his business decisions in the finest detail, the ultimate decision is based on his 'gut feeling' about whether an opportunity will work out or not. Kahneman argues that this type of decision-making is not based on randomness or luck but on intuitive heuristics (rules of thumb) which people tend to develop as a way of simplifying complex multidimensional decisions. In other words, in making a complex decision, we often resort to a much simpler decision instead, without even noticing we have done so. Over more than two decades, Kahneman's work has revealed a wide range of decision heuristics which guide this spontaneous type of thinking, which he calls 'System 1 thinking'. But he also acknowledges that this spontaneous and automatic approach to making decisions occasionally fails and that people are forced to switch back into a slower, more rational and effortful mode of thinking which he calls 'System 2 thinking'. For example, consider the following question:

> *A bat and ball costs $1.10 in total. The bat costs $1 more than the ball. How much does the ball cost?*

Most people quickly answer 10c because it seems obvious. However, if the ball costs 10c then the bat must cost $1 which is 90c more than the ball. The correct answer is 5c with the bat costing $1 and 5c. Of course, once we have made this mistake, we are unlikely to do it again. Once we know the trap of being emotional and impulsive, we tend to learn quickly and use the System 2 approach to be much more conscious, orderly, logical and careful.

Despite these limitations of System 1 thinking, Kahneman's work shows the importance of trusting one's intuition and trusting the rich and detailed model of the world which is maintained in the associative memory. Kahneman notes that our memory is a vast reservoir of knowledge which we have acquired over a lifetime of experience, practice and learning which automatically and swiftly produces judgements and choices which on most occasions are accurate. The problem, he argues, is that it is very difficult to see when situations move outside our frame of expertise and experience, when our intuition is wrong and when we need to invoke our System 2 thinking to subject it to further analysis.

Non-linear thinking

Cognitive research shows that creative people's brains generally do not work well in a linear plane but work by looking for patterns, connections, associations and things that appeal to their interests and senses. Even for 'normal' people, research has shown conclusively that people remember smells, pictures, highlighted text, coloured maps and diagrams far better than they remember pages of featureless black and white text. In one famous experiment by Haber (1970) people presented with 2,500 images which were stimulating to the senses remembered between 85 per cent and 95 per cent of them one hour later. Other experiments have shown far higher recognition rates up to 99 per cent when combined with other senses such as touch, taste and smell.

What happens in the brain when people are presented with such stimuli has been referred to as 'radiant thinking' because each bit of information radiates millions of cognitive hooks which creates a gigantic cognitive map of associations. The creativity technique of mind mapping has been developed as a 'brain compatible' thinking tool to represent these cognitive patterns. The beauty of mind mapping is that it combines thinking from both the right and left hemispheres of the brain. Our understanding of how these two hemispheres work was developed in the 1970s by Dr Roger Sperry, who won a Nobel Prize for his research. He found that the 'thinking cap' of our brain called the cerebral cortex is divided into two major hemispheres (left and right) which performed distinct intellectual tasks. People who have a dominant right side tend to be more radiant in their thinking which makes them more creative, more imaginative and better at seeing the bigger picture (gestalt) of connections between different knowledge domains. In contrast, dominant left brained people are more linear, analytical and logical in their thinking.

Learning – the trap of distinctive competence

Most managers would agree that the concept of learning is central to an organization's ability to innovate. The concept of the 'learning organization' was first coined by Peter Senge in his seminal book *The Fifth Discipline* (Senge 1990). Senge's work became so influential that he was named by the *Journal of Business Strategy* as the 'Strategist of the Century'. Senge argued that learning organizations are able to excel in their market because they are able to innovate and adapt to their environment faster than their competitors. As a result of Senge's work, the concept of learning has been widely accepted and embraced as being a positive attribute in the field of management, innovation and organizational studies.

Senge's work is now over twenty years old and not surprisingly is widely seen as outdated (although many firms continue to pursue the idea in ignorance of these less publicized criticisms). For example, March (1999) has argued that the whole process of learning as conceptualized by Senge (1990) can be damaging to innovation because it can lead to what he described as 'strategic myopia'. March argued that by facilitating positive feedback between successful experiences and organizational competence, learning processes automatically reinforce a focus on limited areas of successful business which encourages greater and greater specialization in those areas. This in turn encourages organizations to build 'distinctive competences' around these areas of assured success which will more likely yield immediate advantage over competitors. However, as March points out, strategies that are good for the short term are not always good for the long term and the problem with this process is that learners become increasingly removed from other bases of experience and knowledge which might signal broader changes in their environment in the future. Furthermore, an increasing focus on a narrow set of capabilities reduces the capacity to adapt and change capabilities in responding to different futures. In other words, short-term learning that creates a strategic advantage in one domain can be a liability in the longer term by creating organizational rigidity and a tendency to coalesce around existing known capabilities at the expense of developing new and dynamic capabilities to adapt and innovate for the future. March (1999: 204) calls this cycle the 'trap of distinctive competence'.

More recent research has suggested that a capacity for 'unlearning' may be more important than learning in today's increasingly uncertain and dynamic world. The idea is businesses can be become constrained by 'path dependencies' as a result of becoming locked into past routines. These ideas are incorporated into contemporary learning theories through the idea of triple loop learning

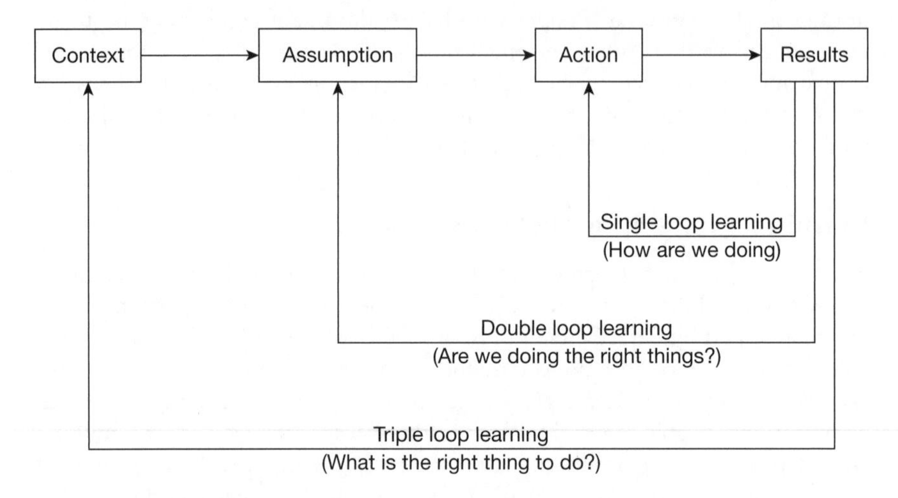

Figure 6.1 The differences between single loop, double loop and triple loop learning (Adapted from Wang and Ahmed 2003)

which was first conceptualized by Wang and Ahmed (2003). The idea of triple loop learning is illustrated in Figure 6.1 and built on the formative work of Chris Argyris (1999), a Professor Emeritus at Harvard Business School who first pioneered the idea of double loop learning

Single loop learning is the lowest level of learning where we might 'react' to a situation and make small changes to specific practices or behaviours without examining or challenging our underlying reasons behind them.

Double loop learning is a deeper form of learning which addresses the causes of a problem and involves changing the underlying practices and procedures which lead to it.

Triple loop learning is the deepest form of learning which goes beyond causes to address the fundamental principles which might have led to those causes taking place. This involves deepening our comprehension of 'why' we chose to do things we do and changing our mental models. These are assumptions and memories held by individuals and organizations which underlie what they do. In other words, unwanted values need to be discarded in a process called 'unlearning'.

The role of technology in innovation

Relatively speaking, construction remains a relatively low-technology, labour-intensive, craft-based industry, in comparison to other industries such as manufacturing. There are many obvious reasons for this. The bespoke nature of each project and the relatively uncontrollable production environment in which construction occurs does not lend itself to mass production methods or technologies. Furthermore, resistance to technological change has been institutionalized by traditional ways of working which go back hundreds of years. And as we have already pointed out, many of these practices and power structures continue to be protected by powerful professional and trade associations.

While there may be significant barriers to the adoption of technology in the construction sector, new technologies undoubtedly have the potential to transform the industry. As we show in chapter five, where specific technologies are discussed in detail, some sections of the industry are successfully using new offsite fabrication technologies to significantly reduce accidents, improve product quality and reliability and increase the speed of construction. Advanced simulation and communication technologies are also enabling businesses to test new ideas at lower costs and faster speeds than ever before. And many companies are experimenting with sophisticated materials tracking systems which can manage their logistics and supply chains more effectively. In particular, new technologies are enabling large geographically dispersed businesses to capitalize on their natural economies of scale without sacrificing the flexibility and responsiveness of a small business. As we pointed out in chapter one, this is leading to a new fifth generation of innovation where large firms are driving innovation in response to many of the world's most enduring challenges such as climate change, environmental degradation and poverty.

While technology offers many potential benefits to firms, particularly large firms, leveraging these benefits is a challenging process and the implementation and adoption of technologies should not be taken lightly. Research shows that those companies that benefit the most from new technology know why they are using it, know how to use it and know how to adapt their business to leverage its benefits. The social reality of technology implementation is highly complex and research shows that technology's impact on business performance is contingent on a broad set of factors. Liker et al.'s (1999) research shows that these include:

1. The reasons for its introduction
2. Management philosophy and organizational structure
3. The labour-management contract
4. The degree of a shared agreement about technology
5. The impact on work organization
6. The process of technology development
7. How change associated with technology is managed in terms of industrial relations, skills, training, structures and systems etc.

New technologies will significantly challenge traditional power structures and introduce new tensions into working relationships and supply chains as traditional functional boundaries are challenged and change. As Dodgson et al. (2005) argue, this means that new technologies will require firms to reassess and more clearly define their role in the new value streams that emerge. In traditional industries like construction, this will involve developing a new generation of professionals with multi-disciplinary and technological competencies who are able to work across traditional professional boundaries, relationships and knowledge domains. Furthermore, as technology rapidly changes and reduces in cost, success will depend on shifting strategy away from the ownership and possession of technological resources to focusing on the 'absorptive capacities' which will enable firms to integrate and use the new technologies more effectively than their rivals.

The role of research in innovation

Despite the limitations of traditional R&D models of innovation in industries like construction which are dominated by small firms, R&D investment continues to be an important policy instrument and international measure for innovation activity. Consequently, as we pointed out in chapter one, construction is continually criticized for being a low innovation sector, largely on the basis that its R&D investments are lower than those of other industries. While we have argued that much of the innovation in the industry is of the 'hidden' kind, driven by the need to solve complex problems on a day-to-day basis, does this mean that there is no role for R&D in the sector?

Traditionally, R&D is seen as a knowledge creating process underpinned by rigorous scientific investigation which leads to the commercial development of new services and/or products. R&D can be *applied* or *pure*, the former being a response to market developments and having a practical application, the latter being more conceptual and exploratory with the aim of adding to our knowledge base without any specific application.

Pure R&D has traditionally been the responsibility of the public sector and applied R&D the domain of a relatively small number of research-intensive industries like pharmaceuticals, defence and electronics. Government support for pure R&D has always been considered crucial because research shows that most private companies, if left to their own devices, would under-invest in this area. The private sector tends to see pure R&D as being too lengthy a process and often experiences difficulties in capturing its benefits. Furthermore, it is easy for competitors to access the intellectual property produced which is often difficult to protect. Indeed, even applied R&D can be daunting for firms to sponsor. For example, Kevlar took eighteen years to develop and its commercialization took seven years. Similarly, Xerox took over fourteen years and many setbacks to persuade industry about the value of its new photocopier technology in replacing traditional manual printing machines. As Hilmer and Field (2011: 63) point out, 'brilliant ideas can't be rushed' and many research projects which have shown little social or commercial value in their first decade have led to remarkable outcomes in their second decade. Hilmer and Field's criticisms were not aimed at the private sector but squarely at governments whose attitudes towards research funding are increasingly reflecting private sector sentiments. Hilmer and Field argue that in the quest for greater accountability and instant success governments are failing to satisfy their responsibilities to invest in long-term research programmes that the private sector is unlikely to support. Instead they are favouring more politically acceptable three-year rolling competitive research programmes which often bring existing research programmes to a premature end. These views are supported by Matthews (2008) and Cambourne (2007) who point out that Albert Einstein's 'Theory of relativity' and Charles Darwin's 'The origin of the species' could never have been completed in the present commerciality driven government research funding environment.

How relevant is this debate to construction R&D? How relevant is blue-sky R&D to construction? And with dwindling government support for such research, who is left to fund this type of long-term thinking which is essential to the future prosperity of all industries?

Certainly, it is highly unlikely that the small firms which dominate the construction sector will ever have a significant role to play in this type of long-term research. Instead, as Gann (2000) points out, it is much more likely that innovation will continue to take place 'around' the construction industry rather than 'within' it, meaning that the absorptive capacity of firms to adopt new ideas will be the critical factor in driving innovation in the sector. While this may be true, the problem with this approach is that firms outside the industry will be unlikely to form a much needed vision for the industry as a whole. This must fall to the relatively small number of large firms that dominate the industry in turnover. But how do large firms in the construction industry do this effectively, when they rarely collaborate, when attitudes towards R&D are so negative and when they do not have the traditions of managing R&D and the dedicated resources for doing so, that other industries invest? In seeking an answer to these difficult questions the construction sector might be able to learn from other industries. For example, Braben's (2004) detailed case study of British Petroleum's (BP) highly successful Venture Research Unit might have some potential answers.

BP's Venture Research Unit was deliberately located outside any of BP's existing business units to enable it to generate 'new unimaginable breakthroughs and ideas' that would lead to new industries and markets for BP. Working under the management of BP's Venture Research Advisory Council and in close collaboration with the world's leading universities, BP's innovation strategy involved signing up the world's most gifted pioneering scientists whose interests were aligned with the business. Rather than following the traditional approaches to R&D which typically involves commissioning pre-determined business-led projects, BP provided these top academics with the resources to pursue their own ideas and to launch radical challenges to existing ideas outside any external influences and constraints. This process not only preserved the ideology of independent unbiased research but was designed to promote uninhibited thinking. BP chose academics on the basis of whether their research would radically change thinking about something that was very important to society and to BP's business. Once accepted into the Venture Research Unit's team, BP's goal was to help these leading academics bring this about. There was an exceptional lightness of touch in managing this research. There were no formal contracts, no formal objectives to measure progress and no outcomes specified. The only requirement imposed on the academic team was that they were asked to keep BP regularly informed of what they were doing so that BP could be the first to translate these ideas into marketable products and services for their customers. Academics were not concerned directly with the commercialization process. This was entirely BP's responsibility and once an academic received BP's money, they were free to use it in any way they liked. BP did not dictate projects, fields of study, problems or timescales. Eventually the unit's funding was expanded to a consortium of business partners with complementary interests in BP's demand and supply chain such as ICI, Sony and DuPont. While a very small fraction of the academic output was eventually taken up by BP, those ideas proved to be extremely successful and their return-on-investment more than covered the relatively small investment in the overall initiative.

The beauty of BP's collaborative approach was that it avoided the classic problem of selecting research proposals. By supporting individual leading researchers and their research aspirations rather than specific research projects, BP was able to pursue a liberal approach which drew knowledge from a range of disciplines and business partners. Furthermore, by hand-choosing their research partners and by minimizing the normal time, resource and bureaucratic constraints associated with scientific research, BP not only reduced the barriers to innovation but they also reduced risk since the researchers they supported were almost certain to succeed. The key question and risk then became how to convert that research into ideas for BP's benefit.

BP's research strategy was innovative but was not completely new. For example, Siemens had pursued a similar strategy since 1869, General Electric and Bell Telephone Company since 1900, DuPont since the 1920s and IBM since the 1960s. Each of these company's R&D programmes were renowned for their creativity, freedom and transformational innovations. For example, IBM set up the 'IBM Fellows' programme where key employees were appointed as dreamers, mavericks and heretics (wild ducks) with the freedom to think outside the square and call on any part of the company they needed to do so. The task of IBM's Fellows was to 'shake up' the system and some of their successes (such as the scanning tunnelling microscope) led to Nobel Prizes for the IBM Fellows concerned. DuPont's R&D program was also renowned for its creativity and led to innovations such as the development of nylon. And the Bell Telephone company's R&D programme produced no less than eleven Nobel Prize winners for innovations like the development of communication satellites and the detection of the faint background noise from the universe's creation which eventually led to the Big Bang theory. As Townes (1999) points out, these innovations happened because scientists were enabled to freely exchange their ideas in safe independent environments free from regulation, political influence and vested interests. Few of these discoveries were 'needed' before worked started on them, many were unplanned and none were guided by the strategic priorities of firms or government research panels.

While there are precious few examples like the above in the construction sector, there are of course exceptions to this rule. For example, the engineering firm Arup, like IBM, has appointed senior business Fellows to drive future thinking in their business and invests more in R&D than any other firm (approximately 3 per cent of annual turnover). Arup has a clear road map for research which extends over twenty years into the future based on key drivers of change in key business areas. These roadmaps provide a strategic focus for its research and help to inform priorities for funding, reveal new market opportunities, identify client needs and business directions and plan collaborative partnerships. Arup also integrates research-based KPIs into performance reviews for staff who are required to publish and collaborate with universities in creating new knowledge for the benefit of the business and society at large. Arup's research programme is driven by 'pull' from its business leaders and 'push' from universities and its research network partners. Research is seen as vital for its strategic business planning to ensure that it is equipped for future trends and that it can capitalize on opportunities to improve its business and enter new markets. Arup invests in many research projects globally, and its research funding is deployed in a strategic and coherent fashion in order to deliver results for the business. Projects are typically a mix of internal and external R&D, with a particular focus on supporting collaborative, applied academic research. As in BP's case, Arup's research team's work involves forging and maintaining links with the best quality universities, researchers and research establishments – no matter where they are. Arup also works closely with government research funding and advisory bodies around the world to lobby for certain priority areas of funding and to leverage its own resources. To ensure Arup capitalizes on the Intellectual Property inherent in its research, it has also established an IP Executive (IPX) group. As a result of these collective initiatives, Arup has arguably generated more knowledge than any other firm in the construction industry and has benefited enormously as a result. One example of a research driven innovation which has arisen out of this programme is the award-winning self-installing platform for sea gas developments which has delivered over £20m in fees.

While there are undoubted benefits for firms in having a formal R&D programme, the era of large-scale R&D driven innovation is coming to an end and there are a number of important reasons for this:

1. It encourages the view that innovation needs to arise out of some large-scale and costly laboratory-based research programme which is restricted to large firms.
2. It ignores the role of individual entrepreneurs who play a critical role in driving innovation and economic growth.
3. It underplays the importance of serendipity, human creativity and imagination in the innovation process.
4. It ignores the capacity of communication technologies and social media to drive innovation in any business no matter how small.
5. It does not reflect the increasingly connected society in which we live and the increasingly complex multidisciplinary challenges we face, which require collaborative inter-firm and inter-sector alliances to address.

Indeed, the claimed benefits of traditional large R&D programmes on productivity, growth, turnover and profitability are highly contentious at both a national and individual firm level. Some studies show a positive relationship but many show a negative or no relationship at all (Masson et al. 2010). For example, Baumol (2010) argues that the inherent conservatism and planned nature of traditional R&D programmes can cause large firms to avoid taking the risks and imaginative leaps of faith that all revolutionary breakthroughs require. Furthermore, structured R&D activities can 'crowd out' the individual innovative entrepreneurs who have been responsible for the bulk of revolutionary breakthroughs in society. Numerous studies have shown that it is small enterprises that are the catalyst for most of these transformational innovations and that small firm innovation is significantly more leading edge than innovation from large firms. Baumol (2010) shows that small firms are thirteen times more innovative per employee than large firms and are twice as likely to be listed among the top 1 per cent of highest impact patents. While it is large firms that often commercialize new ideas, the majority of them originate in small firms who play a critically important role in up-front idea generation.

Given that the construction industry is dominated by small firms, there would seem to be considerable potential for innovation in the sector. However, as we showed in chapter five, there are many cultural, organizational and institutional barriers in the industry which prevent this from happening. The challenge for large firms in the industry is to develop new ways to harness the untapped productive and innovative potential which lies untapped in their extensive supply chains.

The role of universities in innovation

Palangkaraya et al.'s (2010) research into the sources of knowledge used by Australia's most innovative firms in the resources, manufacturing and services sectors revealed a disturbing fact for university managers. Universities ranked last in importance out of six key sources of knowledge behind business competitors, conferences, customers, public sector bodies, patent literature and suppliers. These findings are supported by research in the construction industry by Carthey (2007) who found that the main sources of knowledge for leading architectural practices in Australia was the internet followed by trade journals with academic peer reviewed or academic studies ranking very lowly. Carthey found that in solving day-to-day design problems, architects preferred to talk to each other and rely on their own experience from past projects rather than to read about the latest academic research.

Although universities and the academics that inhabit them often have a tenuous, suspicious and even tense relationship with industry, the knowledge they produce has become a critical resource in today's increasingly knowledge-driven global economy. This has been reflected in the growth of the university sector over the last two decades. For example, Australia's international education industry has grown enormously in recent decades and is now Australia's fourth largest exporter behind coal, tourism and iron ore. In simple terms, the role of universities is to create new knowledge and impart it to people for the benefit of the community's wealth, health, security and well-being. They do this in two main ways. First, they search for new 'fundamental knowledge' which business is unlikely to invest in, but which often plays a crucial role in the early gestation phases of the business innovation process. Second, they play a central role in creating the human capital that is so crucial to innovation by educating people, opening their minds and giving them the skills to generate and implement new ideas. As Cutler (2008: 45) points out,

> Innovation is fundamentally a people-driven exercise and any nation's capacity to innovate is inextricably linked to the breadth and quality and focus of its education and training system . . . A highly skilled workforce is essential not only for the generation and application of new knowledge, but also to use and adapt knowledge produced elsewhere.

However, the role of universities in society has changed fundamentally over the last hundred years. Up to the middle of the last century, higher education was expected to provide basic and professional education, to discover new knowledge and to preserve past knowledge. As Katz (2002) points out, knowledge grew slowly and incrementally and society was happy to leave this to universities. But all this changed during the early part of the twentieth century when the discovery of atomic science and the Cold War led to the creation of national science foundations in many countries, accelerating the pace of scientific research and creating what we now recognize as the modern 'research-intensive university'. While universities in most countries remained largely publically funded during this expansionary period, competition for knowledge from business caused money to flow into universities, accelerating the growth in university infrastructure. While knowledge was generated at an incredible rate during this 'golden period' of unprecedented university growth, there was also an unintended side effect, a downside which we are only now starting to recognize. Before this period of unprecedented growth, universities were quite liberal environments which tolerated many types of thinking and knowledge. But as universities grew and competed for a greater part of the enormous funding pie, many academics were forced to fit into the highly fundable 'scientific' model of knowledge creation which was increasingly perceived by Western society to be associated with rapidly growing post-war prosperity.

The importance of research funding as a measure of university and individual success led to the rapid growth of every field of research which could utilize the scientific model (including construction). While this produced increasing amounts of knowledge in these areas, the downside was increasing knowledge specialization as growing numbers of academics focused on smaller-and-smaller problems to advance the knowledge in their chosen field. This in turn led to significant narrowing of education as increasingly specialized academics confined their teaching to their own specific research fields. Furthermore, since there was little credibility or reward for doing anything else other than publishing, few academics would spend time innovating to construct curriculums to help students connect and make sense of the highly specialized knowledge they were producing. The dominance of publications as the main criteria for hiring and promoting academics also led to a proliferation of academic 'peer-reviewed' journals which had grown to communicate the vast amounts of data that was being produced. And as academia became more professionalized, the methods and language of academics also developed, becoming largely indecipherable to the outside world.

This trend has continued unabated. Today every aspiring university wants to be 'research-intensive' and many business leaders complain that research has become increasingly divorced from the needs of industry and broader society. While academics want to know more and more about less and less, politicians and practitioners needed to know less about more and more. As Locke and Spender (2001) point out, business and governments tend to focus on big-picture issues and have great difficulty understanding the value of the highly specialized knowledge which many academics produce. Bennis and O'Toole (2005) concur and argue that business schools are institutionalizing their own irrelevance by separating management decision-making from the practical, social and ethical context in which it occurs. And as a result of these growing concerns, it is not-surprising that the national political and industry atmosphere in many countries has become hostile to universities and academics.

The construction sector's attitudes towards university-based research mirrors this trend. Privately funded R&D in the construction sector remains very low compared to other industries. And as Lenard's (1996) and Loosemore et al.'s (2002) research found, most people in the construction sector perceive research to be an academic activity which in most instances produces theoretical results of little commercial value. Loosemore et al.'s (2002) survey of attitudes towards innovation in the construction sector reported that 90 per cent of those interviewed considered that R&D produced an unsatisfactory rate-of-return and 60 per cent of respondents did not believe that R&D enhanced their competitiveness. Many of the respondents who had sponsored research had been disappointed with their experience and perceived a wide gulf to exist between the culture of researchers and industry. To most practitioners, academics produce more problems than answers and their proposed solutions are, more often than not, too radical to implement. Indeed, most practitioners felt uncomfortable interacting with academics and had a strong preference to commission consultants to undertake their R&D or to undertake their own in-house research. In essence, academic R&D was perceived to provide poor value-for-money.

There is no doubt that this divide between the university and business sectors is potentially damaging to the construction industry. While management consultants provide more acceptable answers, more rapidly and in a much more palatable way, it is critical that firms base their business strategies on scientifically derived evidence generated in an independent unbiased environment. It is also critical that industry continues to invest in the type of 'blue-sky' long-term thinking which is undertaken within universities. Throughout history, there are literally thousands of examples of seemingly abstract ideas which have led to important practical applications after many years of painstaking research. Take for example the development of nuclear energy which spanned over forty years of research beginning with Henri Becquerel's discovery of radioactivity in 1896, Ernest Rutherford's discovery of atomic structure in 1911, culminating in the development of quantum physics and experiments performed in Berlin by Otto Hahn in 1938. Even more patience was required to see the benefit from solving the '12-coin problem' which baffled mathematicians for almost a century. The problem was to identify one counterfeit coin in a group of twelve with no more than three uses of a balance scale. Few would have predicted that the solution, which was to weigh groups of coins, would become the basis of group testing which was used during the Second World War to screen recruits for syphilis without having to test everyone. More recently, who would have thought that Barbara McClintock's study of maize over twenty years ago, which found that some genes can control others, would lead to modern industrial developments in genetic engineering, immunity medicines, cancer drugs and antibodies. And it is not just research in the physical sciences which can provide longer-term insights which can benefit business and society. For example, anthropological research by the late Clyde Mitchell from Oxford University into kinship ties in African villages during the 1950s led to the development of social network analysis techniques which have become widely used in business to understand and optimize communications in complex organizations. In recent years social network analysis has also been used by governments around the world to understand the spread of radicalism and terrorism, infectious diseases such as Aids and other developments such as social media technologies.

To reverse the irrelevance to, and the growing separation of, universities from business and society, Dodgson (2008: 31) argues that universities have to find new novel ways of partnering with business, governments and communities. For example, NESTA's (2009) report entitled *The Connected University* proposes a new model which involves the development of regional clusters of excellence around regional networks of firms with national and international connections. It is also clear that universities need to recruit more broadly. Rather than filling their faculties with young inexperienced researchers who have had no exposure to industry, universities need to recruit entrepreneurs with experience in the private and public sectors who can see the commercial value in new research and broker connections with firms that could benefit from them. Universities also need to ensure that their technology-transfer arms better capture, harness and diffuse the knowledge generated by their staff. This knowledge should not be just scientific, technical and engineering orientated but should also draw from the humanities and social and behavioural sciences. Finally universities should recognize that time spent in industry makes an important contribution to academic advancement and they should make more information available to researchers about options for secondments to work in the private and public sectors.

The role of government in innovation

Given that many private firms are not prepared to take the risk involved in blue-sky academic research and will wait for the results to eventuate, the role of government is critical in the innovation process. Indeed, many of the problems we face today, such as climate change, are so complex that they can only be solved at an international level, through international governmental organizations such as the G20, ASEAN, UN and EU.

In general however, outside the international context, any government's role in driving domestic innovation requires them to do ten main things:

1. To step-in to fund the basic research that private firms are not likely to support.
2. To assist entrepreneurs to commercialize research by ensuring sufficient mentoring, support and venture capital is available to fund start-up companies.
3. To reduce the risk of innovation by providing a stable and conducive economic, legal and regulatory environment for it to occur and to cultivate confidence in capital markets.
4. To provide leadership by insisting on innovation through their purchasing power and procurement arrangements as a major client.
5. To stimulate innovation through incentives, competition and innovation policies.
6. To build up a country's human capital through their education systems and immigration policies which provide a talented pool of human resources to provide innovative ideas.
7. To take an overall view of a national innovation system by developing a national innovation strategy.
8. To develop a country's social and market capital by brokering connections between the various internal and external elements needed to make it work, connecting business into new global markets.
9. To create a positive and stable regulatory environment within which innovation can occur.
10. To reduce red tape and simplify regulations procedures and requirements.

While there is a whole debate about the role of governments in innovation from interventionist to laissez-faire, there is a view that in industries like construction, regulation has an especially important role in driving reform. In their interviews with over thirty thought leaders in the UK construction sector, Loosemore and Holliday (2012) argued that while there is evidence that too much regulation can stifle innovation through red and green tape, there is also little history of voluntary change within the industry. Many thought leaders therefore promote the view that many parts of the industry need to be encouraged to innovate through regulatory reform. In particular, innovative firms in the construction sector are often disadvantaged because they are undercut by those firms which do the minimum and deliver standard solutions. While much of this problem is attributable to unimaginative clients who do not have the interest, inclination or tools to value innovation, regulation can also help by setting minimum standards which make it difficult for the industry's 'long tail' of underperforming firms to operate.

The whole debate about government's role in innovation is too enormous to review here. Nevertheless, it is clear that to drive innovation, firms need to actively lobby governments. And, despite high levels of competitiveness in the industry, it is far more effective if they do this together rather than in isolation.

The ethics of innovation

At the very start of this book we exposed the pro-innovation bias which afflicts the innovation literature and argued that it is crucial to consider the varied perspectives and impacts (both positive and negative) which innovation can bring about. For example, the development of green buildings will reduce our reliance on fossil fuels and reduce environmental pollution but will also inevitably mean redundancies in some sectors of the construction industry which have built their businesses around traditional fossil fuels. Similarly, while investments in green energy sources like nuclear, wave, wind and solar promise to reduce greenhouse gas emissions and spawn new clean energy industries, they also raise many concerns relating to the safety and well-being of communities and the destruction of natural habit. So with every stage of the innovation process, from the creation of the original idea through to its implementation, come many ethical questions which must be dealt with to the satisfaction of society. An understanding of ethics is therefore crucial to being able to implement innovation in practice.

Ethics is a branch of philosophy which, at its simplest, can be described as a system of moral principles which affect how people make decisions and lead their lives. The field of ethics provides many frameworks for making and evaluating business decisions, particularly in ambiguous situations where moral values and organizational rules clash (Brown 2005). At the heart of ethics are our belief about what is right or wrong and a concern about something or someone other than us. These ideas which develop over many generations in our communities, families, religions, philosophies, cultures and institutions are continually changing and are socialized through education and our membership of various organizations, community groups and societies. The ethical principles imparted to us by these constituents offer us ethical rules, moral maps and principles that we can then use to find our way through difficult issues.

Philosophers have long contemplated the subject of ethics and Table 6.4 describes the basic principles of some common ethical schools of thought which have emerged.

Table 6.4 shows just how easy it is for people to construct an ethical argument for or against any new idea they may have. It also shows how ethical codes can easily clash in making judgements about the morality of innovations, and that

Table 6.4 Common ethical schools of thought

School of thought	*Basic principles*
Supernaturalism	Supernaturalism draws its principles from religion. It teaches that the only source of moral rules is God and that an innovation is good because God says it is.
Intuitionists	Intuitionists think that goodness does not need justifying or proving. They think that basic moral truths of what is a good or bad innovation are self-evident to a person who directs their mind towards moral issues.
Consequentialism or utilitarianism	Consequentialism or utilitarianism bases morality on the consequences of innovations and not on the actions of innovators themselves. So innovators should do whatever produces the greatest good or 'utility' for the greatest number of people.
Non-consequentialism or deontological ethics	Non-consequentialism or deontological ethics is concerned with the actions of innovators themselves and not with the consequences of innovation. In other words, an ethical innovation is one that is intended to do good whether or not good comes of it or not, assuming it comes from a sense of duty rather than self-interest. Associated with the writings of Immanuel Kant who set out a universal set of moral principles for ethical behaviour, deontological (duty-based) theories are about 'doing the right thing'.
Virtue ethics	Virtue ethics emphasizes the character of the innovation agent, rather than their intentions or consequences, as the key element of ethical thinking. To put it very simply, virtue ethics teaches that an innovation is right if and only if it is an action that a virtuous person would do in the same circumstances, and that a virtuous person is someone who has a particularly good character. Although concern for virtue appears in several philosophical traditions, the roots of the tradition lie in the work of Plato, Aristotle and ancient Greek philosophy.
Situation ethics	Situation ethics rejects prescriptive rules and argues that individual ethical decisions should be made according to the unique situation. Rather than following rules the innovator should desire to seek the best for the people involved. There are no moral rules or rights – each case is unique and deserves a unique solution.
Ideology ethics	Ideology ethics argues that ethics is the codification of political ideology, and that the function of ethics is to state, enforce and preserve particular political beliefs. It argues that innovations are unethical if they are used by the dominant political elite to control everyone else – but particularly if they do not apply this code to their own behaviour.

an understanding of ethics is not likely to provide a definitive single right or wrong answer to moral problems associated with an innovation. Indeed, the field of ethics does not purport to do so. Rather, it simply purports to provide us with a set of principles that can be applied to particular cases to give those involved some clear choices. Some philosophers go further and would argue that all ethics can do is eliminate confusion and clarify the issues surrounding an innovation. After that it is up to each individual stakeholder to arrive at their own conclusions.

Conclusion

The aim of this chapter was to explore the organizational factors that can help firms in the construction sector better harness the innovative potential which lies untapped in the dynamic, interconnected and uncertain business world in which they now exist. We have done this by stepping outside construction into the field of philosophy, psychology and the social and behavioural sciences. We have learnt that the key to innovation is not to follow the crowd but to foster curiosity, imagination, nonconformity and a capacity to think which draws on a liberal knowledge-base and an ability to adapt and cope with change. This in turn depends on creating a flexible, dynamic and collaborative organizational structure and culture which empowers people to think creatively and which tolerates ambiguity, risk and failure. This last point, which highlights the centrality of risk to the innovation process, is a nice segue into the next chapter which discusses how the significant risks associated with the innovation process can be effectively managed and turned to advantage.

7 Managing the risks of innovation

I have not failed. I've just found 10,000 ways that won't work.

Thomas Edison

Introduction

Innovation is an inherently risky process. There is plenty of empirical and anecdotal evidence to illustrate that innovation is associated as much, if not more, with failure than it is with success. Not only does the act of innovation involve making decisions which may turn out to be wrong, but in the never-ending cycle of creative destruction, new products, services and technologies fail and become outdated. These are then replaced by other products, services and technologies, which in turn fail and become outdated. So being able to manage risk effectively is an essential part of the innovation process.

In discussing how this can be done, the intention is not to present a detailed technical exposition of the risk management process because there are plenty of excellent books on this subject. Rather, the aim of this chapter is to develop a better understanding of the risks which are likely to be encountered during the innovation process and to provide new insights into how to turn them to advantage. Pursuing risky innovations without understanding the risks involved or a firm's capacity to manage them is one of the main reasons why many new ideas fail.

The risks of innovation

Although every innovation will face a unique array of risks and opportunities, Dodgson and Gann (2010) discuss the main risks which can undermine the innovation process. Given the discussions in previous chapters, we can now expand this significantly and the result is the generic list of risks and key questions in Table 7.1.

Table 7.1 The risks of innovation

Risk	*Key questions*
Market	Is the market ready for the new service or product? How far ahead of the market is the new idea? How do we influence the market to accept our new idea? What are our competitors doing?
Strategy	Do we have the business strategy to drive and support innovation? Does it make innovation a priority? Does it provide a focus for innovation around which people can coalesce? Does it provide adequate resources to support innovation? Can our business strategy adapt to changes in our business environment? Was it built top-down or bottom-up?
Leadership	Do our leaders provide the leadership to inspire, encourage and enable people to innovate? What attributes do our leaders have (are they visionary, inquisitive, open-minded and tolerant of experimentation and failure)? Are our leaders prepared to back-up the development of new ideas with appropriate resources?
Client	Do our clients and customers value innovation? Do they encourage it? Will they pay for it? Are they willing to play a role in the innovation process? Can they be educated about the benefits of innovation?
Competitor	What are our competitors doing? How do we differentiate ourselves? Will our competitors beat us to it? Will our competitors develop an alternative and better product? Will our competitors copy us and if so, how long will they take?
Finance	How much will innovation cost us? Where will our venture capital come from? What will be the terms of repayment?
Timing	Should we be a first mover or follower?
Technology	If there is new technology involved, how reliable is it? Will we have teething problems? Do we have the skills and resources to support it reliably? How do we integrate the new technology into our business? What roles and processes will need to change? Do our potential customers have compatible technologies needed to use it?
Human resource	To what extent can people in our business get involved in the development of new ideas to take the business forward? Do we have the diversity of knowledge and talent needed to drive innovation? Are we providing people with the resources, time, skills and capacity to innovate? How much will people resist change and how do we overcome with this?
Supply chain	Are our business partners capable of innovating? Do we have equally passionate business partners? Is our supply chain robust and reliable enough to support the new innovation's development and implementation? Are there compatibility constraints with partners (aspirations, systems and cultures)?
Procurement/ contractual	Do procurements systems and contracts encourage integration, openness and trust or do they encourage fragmentation and confrontation? Do we share the risk and reward of innovation with our business partners?

Table 7.1 continued

Risk	*Key questions*
Regulatory /legal	How do existing laws and regulations affect the success of our innovation and how will potential changes in those laws and regulations affect it? Can we bring about changes in those laws and regulations to create new opportunities to innovate?
Organizational	Do we have the business environment, culture, systems, processes, incentives and structures in place to drive, diffuse and support innovation? Is there a sharing of ideas across projects?
Political	Will a change in government or government policy affect the success of our innovation? Are there political ideologies which may prevent the uptake of an innovation?
Institutional	Are there institutional barriers to the adoption of an innovation (unions, professions, associations, governments etc.)?
Economic	Will a change in economic policies or fortunes affect our ability to develop and implement our innovation successfully?
Natural	Will changes in the environment, acts of god, extreme weather etc. affect our innovation?
Social	Will changes in social habits or preferences affect the demand for our new idea?
Security	How can we protect our IP? Can security issues affect the demand for our innovation?

Risk appetite

Given the inherently risky nature of the innovation process, an 'appetite' for risk is an essential prerequisite to successful innovation. In simple terms, an organization's risk appetite defines the extent to which its board of directors is prepared to take the types of risk listed in Table 7.1. Statements of risk appetite are important in setting the boundaries for management decision-making throughout an organization. Without this, an organization is prone to its managers making risky decisions which fall outside its capacity to manage them.

Every organization has a different risk appetite which should be clearly defined and communicated by the board of directors. Some boards of directors are risk averse and prefer their staff to avoid risky activities. Other are risk takers and are more willing to engage in risky ventures. Once an organization's risk appetite is clearly articulated, managers are then in a position to understand what a potential innovation 'means' for its business and whether it is something it should pursue or not. For example, any innovation with a potential risk profile which lies outside the limits set by the board of directors would normally be automatically rejected or dealt with in a specified way. Conversely, any potential innovation with a risk profile within these limits would be permitted to move forward, subject to continual monitoring to ensure its risk profile does not move outside these limits. As well as setting the tone for consistent decision-making throughout an organization, a clear articulation of risk appetite also forces debate about other parts of the business which are crucial to innovation. For example, new information systems and roles and responsibilities may be required to monitor certain new risks associated with innovations to ensure they do not fall outside acceptable limits.

Defining risk appetite involves a number of simple steps. The first step is to define an organization's objectives (in measurable terms). The second step is to define how much downside and upside variation to those objectives managers are prepared to tolerate. The third step is to define the organization's capability and capacity to manage different types of risks. To assist with this, business characteristics which differentiate effective and ineffective risk management are illustrated in Table 7.2.

Managing stakeholder perceptions of innovation risk

Clearly, not all innovations have the same spill-over impacts on society. Some innovations have only a small impact on a limited number of stakeholders while others are so significant that their 'sphere of influence' has a major impact on a wide range of stakeholders across multiple international borders. These stakeholders might include local communities, business partners, clients, customers, governments, NGOs and lobby groups. Regardless of who is affected by an innovation, one thing is true – stakeholders have never been so informed about the risk of business activities on their lives, never been so educated to understand it and never been so empowered to do something about it. So being able to manage stakeholder perceptions of relative risk associated with an innovation has become a fundamentally important, but largely ignored, aspect of the innovation process.

The potential for wider community interest in the innovation process is particularly acute in the construction industry which has potentially enormous impacts (both positive and negative) on the ecological, social, cultural and economic environments in which people live. There is plenty of evidence to illustrate that if community perceptions of risk associated with construction innovations are not managed effectively then they have the potential to trigger long-standing and acrimonious community protests. And when played out in the media, these protests have the potential to cause significant social, political and economic damage to the interests of all concerned. There are many examples of construction projects being delayed by community protest. For example, in the controversial Lane Cove Tunnel project in Sydney, public perceptions of pollution risk associated with innovative ventilation stacks, which were designed to filter and disperse pollutants away from public spaces, couldn't be allayed even after scientific reports showed that pollution levels were not only under control but were lower than they were before the tunnel was built. Teo's (2008) research into community protest against controversial construction projects showed that this scenario is all too common in the construction industry. According to Teo,

Table 7.2 Business characteristics which differentiate effective and ineffective risk management

Ineffective risk management	*Effective risk management*
A culture of success and invincibility. A belief that failures happen to others and that good management and hard work prevents problems. Problems are suppressed and considered a sign of weakness and managerial failure.	A culture of openness, awareness and sensitivity to the risks of innovation. Failures and problems are not seen as a sign of failure but are tolerated and seen as opportunities for learning and improvement.
Task-orientated cultures which stress the importance of profits over people and other corporate goals. Blind hunger for profits and a belief that high profits justify the taking of high-risks.	The end does not justify the means. The need to make a profit is balanced with recognition of wider social responsibilities to society and the environment. An awareness of the spill-over effects of innovation into these areas.
A mind-set that company size and past successes provide protection from innovation risks.	An appreciation of business vulnerabilities and the dangers of complacency.
A belief that that risk management is someone else's responsibility and that they have the power to offload risks onto other parties, thereby insulating themselves from the uncertainties of innovation.	A culture of collective responsibility for managing the risks of innovation throughout the organization and supply chain. A willingness to share the risks and rewards of innovation in a way which ensures that those who bear risks have the knowledge and resources to control them.
Bureaucratic organizations which follow rigid procedures and punish people who deviate from them.	Flexibility and willingness to 'let-go' of formal, standardized systems and procedures which can become restrictive and counter-productive in taking a new opportunity or exploring a new idea.
A tendency to go-it-alone and work secretly on innovations.	An understanding of risk interdependence through supply and demand chains. Recognition of the need to collaborate to innovate and a willingness to do so.
Risk management systems represent little more than a managerial façade to impress external stakeholders and reassure managers that something is in place to deal with the unexpected, even though they know that they have minimal impact upon day-to-day organizational practices.	Proactive risk management is systematically incorporated into strategic planning for innovation and championed by senior executives so that it is an integral and instinctive aspect of organizational life at all levels.
Organizational priorities and core values are unclear and people have to take it upon themselves to develop new ideas and often have to fight the system to do so.	Senior executives champion, lead and support innovation by providing sufficient resources and clear statements of fundamentally held, core beliefs and attitudes relating to organizational priorities.

(Adapted from Mitroff and Pearson 1993, Pearson et al. 1997 and Pearson and Clair 1998).

the assumption of many in the industry is that community consultation is the responsibility of planners and that community consultation will only serve to spark unnecessary protest and costly disruption. The result is that many in the industry engage in token community consultation and dismiss or disregard community concerns as irrational, uninformed, ignorant or driven by nostalgia (Teo 2008).

While Teo's (2008) groundbreaking research vividly illustrated the cost of upsetting stakeholders, it also showed that with meaningful consultation, communities can be an important ally and source of useful intelligence about the potential risks and opportunities associated with innovation. It is clear that effective and meaningful stakeholder consultation should be a central part of any innovation strategy and should take account of 'normal' people's perceptions of risk. As Friedrich Nietzsche, a German philosopher of the late nineteenth century argued, there is no such thing as facts, only individual interpretations of them. No matter what the impacts of an innovation might be in reality, people will respond to innovation according to their perception of risk and not necessarily according to objective measures. While external stakeholder responses to an innovation may seem irrational and uninformed, scientific and technical assessments of innovation risk by experts are relevant only to the extent that they are integrated into individual perceptions. Objective, actuarial and technical measures of risk, although easier to operationalize, have little meaning if they are separated from the social and behavioural context in which risk is experienced by the public. If it is not then people are fully entitled to respond in a way which may appear to others as inconsistent, irrational, illogical and unreasonable, as long as it is within the law.

To avoid this happening, the language used in communication with stakeholders needs to be clear and easy for them to understand. It must assist understanding and be based on a mutual sense of respect for the positions and perspectives of all parties, not be a medium of dominance disguised by indecipherable technical jargon. This in turn requires that innovators understand who the stakeholders are, what they want to know, what they want to achieve and what their preconceptions might be. Successful innovation should be a multi-way process that is designed to make information about risk accessible to everyone and to promote mutual understanding and consensus. Presenting a *fait accompli* to stakeholders is almost certain to result in objections which will undermine the innovation process or even stop it completely.

Perceptions of innovation risk and opportunity

There has been a considerable amount of research into the factors that shape people's perceptions of risk and opportunity. Tversky and Kahneman (1974), Daniel Kahneman, Amos Tversky and Paul Slovic (1982) and Kasperson et al. (1988) are widely credited as the first to offer a conceptual understanding of this process. Through a range of experiments they showed that risk has a cognitive, emotional and psychological dimension and is perceived in different ways by different people depending upon their backgrounds, personalities, experiences and social and cultural environment. People do not always behave rationally in relation to new ideas but in accordance with a whole range of 'cognitive biases' which can cause people to 'amplify' or 'attenuate' the risk associated with it. So the way we communicate with people about risk and opportunity is crucial to the way that they perceive and respond to it.

There are two broad types of biases which people are susceptible to:

1. Personal biases – associated with people's own psychological makeup, education, experience, culture, beliefs values etc.
2. Reporting biases – which are introduced when people talk or report to each other about a new idea.

Personal bias

Personal bias can arise in a number of ways. For example, Kahneman's 'Prospect Theory' shows that most people are inherently risk averse and fear losses more than they value an equivalent gain. In other words, most people prefer not to lose $100 much more than they would love to gain $100. This means that stakeholders are unlikely to weight gains and losses in the same way that an innovator might and that potential new ideas will only be accepted if their gains significantly outweigh their risks.

We also know that people have a tendency to care more about what happens today than what happens next week. So innovations are more likely to be accepted if they offer immediate benefits. Economists refer to this as 'discounting', a principle which shows that people care less for future costs and benefits than they do for immediate ones. In other words, the immediate gratification we get from purchasing something now (even if we have to borrow money) outweighs the extra costs associated with interest payments we will pay in the long run. The rate at which people discount future consumption in favour of current consumption is called the 'discount rate' which is calculated by using interest rates and inflation rates (to discount future values back to net present values). However, rational economic calculations like this are unlikely to hold much credibility with external stakeholders. They assume that people behave rationally in making purchasing decisions, which in virtually all circumstances is highly unlikely to be true. Furthermore, there is an assumption that everything can be valued easily in the future, which again may not be true. Take for example, the relative costs and benefits of tackling climate change which raises many difficult questions. For example, how do we compare the needs and welfare of people living in the future with the needs and welfare of people living today? And how do we compare the value of jobs that arise today from industries like logging with the value of trees for societies in the future? These types of questions inevitably involve making ethical and values-based judgements about the value of things today compared to the value of things in the future. And the more we value the future, the less we will discount it. In other words, the less we will neglect the future to pay for the present. This means that innovations which are perceived to adversely affect the future interests of society will have to offer higher immediate returns if they are to be accepted.

Another important influence upon people's perceptions of risk is their perception of responsibility for it. People tend to amplify the risks that they feel responsible for. For example, research shows that when a person is legally responsible for an event (such as drink driving) and is liable to be penalized in the court system for doing so, they will see the risks of drink driving as being higher than if they were not legally responsible. But people also feel a 'moral responsibility' to act in particular ways, and although it may often be the case this coincides with their 'legal responsibility', the two do not always coincide. Deciding what if anything is morally obligatory is a principal concern of ethics and is determined largely by the free will of people to decide what is right and wrong, regardless of the law. For example, people might believe that when presented with a proposal to build in their community, they have a moral obligation to protect the environment for future generations. They may also feel that the construction firm putting the proposal has a similar responsibility in response to being granted a licence to do so. So as well as accepting moral responsibilities for themselves, people can impose them on others. In this way, people's feelings of moral obligation can significantly complicate the job of understanding relative perceptions of risk between business and society.

Research has also shown that in assessing the risk of a new idea, the public's perception of 'consequences' is likely to hold more weight than the 'probability' of it happening (Weber and Hilton 1990). This is because most people have difficulty in understanding expressions of probability compared to those of consequences. For example, while the overall level of risk associated with a nuclear power plant may be extremely low (due to a tiny probability of meltdown) the consequences of an accident could be catastrophic (and easily visualized and understood in terms of peoples' lives). Therefore, the risk is judged as very high by the public.

The critical factor that people appear to consider in measuring the size of potential consequences seems to be its immediate impact. People tend to be more sensitive to potential short-term losses than to potential long-term losses (or even gains). For example, Rotov and Baron (1990) found that when people are told that a vaccine will have a potential side effect on their children, even though the risk of death is much lower than the risk of death in the long term, they will refuse to administer it. This choice is explained by the psychological concept of 'anticipated regret' where even desirable decisions are avoided if they could cause immediate regret, however small that may be. It is not difficult to imagine how this bias could affect a community's perception of risk associated with the prospect of an innovative building development in their community, regardless of the benefits it may bring in the long term. Ariely (2009) also argues that since people are naturally risk averse, they instinctively over-estimate the cost and under-estimate the benefit when making a judgement. So any development will have to produce significantly higher benefits if people are to ignore the potential costs in the long term.

In understanding how people perceive risk, we also know that they use rules of thumb (heuristics) when interpreting information, especially when dealing with complex issues. A good example of a rule of thumb is the 80/20 (Pareto) rule which implies that 80 per cent of something's impact is accounted for by 20 per cent of the causes. But it is important to remember that rules of thumb are just that. They are usually anecdotal, are not supported by evidence and should not be followed precisely. Furthermore, rules of thumb are usually based on previous experience and when underlying conditions change, the underlying rule is no longer appropriate.

Research shows that heuristics are not simply a sign of mental laziness but a human survival mechanism for situations when decisions need to be made quickly or when there is simply too much information to absorb and assess. As Herber Simon (1956) found in his famous explorations of 'bounded rationality' in human decision-making, the rationality of individuals in making decisions is constrained by the cognitive limitations of their minds and the finite amount of time they have to make a decision. So when faced with too much information about an innovation they cannot understand, people will reduce problems to simple principles and solutions rather than spending the time and effort to optimize their decision (a process called satisficing). The implication for those in the construction industry in dealing with communities is that they must carefully think about the nature, timing and volume of information they present.

Well-known heuristics which affect decision-making are listed and described in Table 7.3 and it is obvious how they could adversely affect interactions between the construction sector and the communities in which it does business.

Reporting bias

Tversky and Koehler (1994) also showed that the way people perceive risks depends critically on the way information about the future is posed. This is known as the 'framing effect'. For example many studies have shown that people are much more sensitive to information about relative risks rather than absolute risks. This was vividly illustrated in the failed introduction of an innovative contraceptive pill in the US when it was revealed that it 'doubled' the risk of thrombosis. What women were not told was that this risk was from a very low base and remained very low in absolute terms (Berry 2004). It has also been found that a risk might appear to be small when looked at in aggregate but much larger when broken down into its constituent parts. For example, to many people the risk of taking a new contraceptive pill may be quite low. However, if they are presented with a long list of how thrombosis could lead to blood clots and then to strokes and then to paralysis, the risks suddenly appear to be a lot larger.

Interestingly, while one would expect lay-people to be most vulnerable to such biases, Morgan and Keith (1995) illustrated that experts are just as susceptible.

Table 7.3 Innovation heuristics

Heuristic	*Description and implications for innovation*
Anchoring heuristic	People tend to overly rely on a specific piece of information to govern their thought-process. Once the anchor is set, there is a bias toward adjusting or interpreting other information to reflect the 'anchored' information. For example, when estimating the cost of a new design innovation, people will tend to assume the cost is the same as previous designs which were similar.
Availability heuristic	People tend to make choices based on how easily data can be brought to mind. For example, a recent piece of research which claims that green buildings lead to more productive workforces is more likely to hold weight than an article published a year ago which disproved it.
Representativeness heuristic	People tend to ignore long-term trends in data in favour of new data that might discount it. For example, people tend to think that a year of cold weather is disproof of global warming despite thirty years of data to show rising temperatures. This is also called the 'base-rate error'.
Naive diversification heuristic	People tend to seek more variety when forced to make one major decision which will be locked-in for a long time, compared to when they can make a number of sequential decisions with shorter-time implications. For example, construction clients are more likely to award a contractor repeat business if they are allowed to make sequential decisions along the way rather than lock-in an innovation up-front.
Flaw of averages heuristic	People tend to rely on 'deterministic' estimates of the future when in reality the future is 'stochastic' (subject to a range of possible outcomes with different likelihoods). For example, construction clients generally make decisions on the basis of a lump-sum estimate of costs despite a range of possible outcomes with different probabilities.
Escalation of commitment heuristic	People tend to justify increased investment in an idea depending on the amount they have already invested in it. For example, a construction firm is likely to keep investing in a new loss making venture in which many millions have been invested, rather than cut its losses and pull out. Also called the 'sunk-cost' effect.
Affect heuristic	People tend to let first impressions inform a decision, even if subsequent facts weigh against it. For example, if a client has had a bad experience of a prototype innovation then they are less likely to choose that innovation again, even if the price is lower than an alternative idea.
Fluency heuristic	People tend to attribute greater value to choices which can be processed more easily than those which are difficult to process. For example, clients will tend to value simple solutions to their problems more highly than complex solutions, even if the complex solution is the best one.

Table 7.3 continued

Heuristic	*Description and implications for innovation*
Recognition heuristic	People tend to afford familiar recognized objects higher value than new objects. For example, when a client has worked with a project team before they are more likely to perceive their new ideas positively.
Effort heuristic	People tend to value an idea on the amount of perceived effort that went into producing it. For example, if someone finds $100 they might go spend it on a whim, but if that $100 is part of their pay cheque, they will not waste it because they have worked for it. From a construction perspective, this shows the potential value of up-front client involvement in innovation such as a new design since it creates a greater sense of ownership and involvement in its creation.
Similarity heuristic	People tend to make judgements based on the similarity between current situations and other situations or prototypes of those situations. For example, if an innovative design has a similar feel to other designs experienced and enjoyed by a client then it will be more highly favoured.
Simulation heuristic	People tend to determine the likelihood of an event based on how easy it is to picture the event mentally. For example, if a client is shown an animated fly-through an innovative design than a traditional two dimensional paper-based perspective, then they are more likely to see it positively.
Social proof heuristic	People who are unsure of the correct way to behave, will often look to others for cues concerning correct behaviour. People tend to believe that someone else's interpretation of an ambiguous situation is more accurate than theirs. For example, an innovation backed up with endorsements it is likely to be adopted more easily.
Familiarity heuristic	People tend to assume that the circumstances underlying past behaviour still hold true for the present situation and that past behaviour can be correctly applied to the new situation. For example, if a client has had a bad experience of an innovation or construction firm in the past then they are likely to assume a bad experience will follow on the next project – even if circumstances are very different.
Peak-end rule heuristic	People tend to judge past experiences on how they ended rather than on how they went. For example, if a client was subjected to a bad experience with one contractor but a worse experience with another (but which ended well), then the client will tend to rate the experience of the second contractor as preferable.
Scarcity heuristic	People tend to value something based on how scarce it is and how easily they may lose it, especially to competitors. For example, a construction client is likely to value an innovative design higher if it creates a unique landmark project or if it is something which is unlikely to offered again or to competitors if they don't choose it.

They proved this by assembling a group of climate experts to discuss the probability that temperatures would rise if CO^2 omissions doubled. After being asked for an expert assessment, the experts were asked to put together a research programme for investigating this effect in detail. Although this gave them no extra information to estimate the probability that temperatures would rise, when they were asked again for their assessment, their estimates rose significantly. Of course, it was only the perception that had changed. The level of real risk had remained exactly the same. Rowe and Wright (2001) who reviewed a large number of studies on expert versus lay person assessments of risk also found no firm evidence to suggest that levels of expertise are related to accurate perceptions of risk. However, they are likely to be different from public perceptions because of emotional detachment of experts and scientists from the world they seek to measure (Slovic et al. 1981). For example, to a scientist, the health risks of building a wind turbine might be measured in decibels. However, public perceptions might be more concerned with the longer-term stress and suffering it may cause. These emotive dimensions of risk do not lend themselves to accurate modelling or calculation and therefore tend to be ignored by experts, scientists and decision-makers. However, they are absolutely central to public perceptions of risk.

Another key factor which influences individual perceptions of risk is the educational limitations of the people assessing it. For example, Gigerenzer (2003) found that 40 per cent of people are unable to interpret correctly even the most basic frequencies, percentages and probabilities. However, they found that many of the difficulties go away if the information is presented in qualitative probabilistic labels such as high, medium or low.

Familiarity with the hazard also influences people's perceptions of risk. For example, there is evidence that people who have become accustomed to living in risky environments naturally underestimate the risks to their safety. Psychologists call this effect 'habituation' and it occurs when repeated stimulation by a risk reduces the fear associated with it. For example, people living near nuclear power stations tend to have a lower perception of nuclear risk than those who live further away. Their repeated exposure to the fear stimulus coupled with the lack of catastrophe gradually weakens the power of the stimulus to evoke a response. Furthermore, people in situations such as this tend to develop coping/survival mechanisms to reduce the risk's impact. The implications of this habituation tendency are important in the diffusion of innovation since it suggests that the lowest risk of rejection will be in communities which have already been exposed to the idea and its associated risks. However, conversely it also warns us that those accustomed to an innovative environment might underestimate the real risks associated with a new innovation and be more likely to fail in its development and implementation. It would also suggest that people from highly innovative environments would have a natural tendency to view themselves as immune from the risks of innovation, seeing others as the primary candidates for failure. This effect is known as 'optimism bias' which has been underpinned by a large body of research which has shown that people who suffer from this are more likely to be egocentric (have difficulty seeing things from another's viewpoint), lacking experience (only had positive experiences of a particular risk) or rely on false stereotypes (seeing specific types of people as vulnerable) (Weinstein 1989). This bias was exposed by Flyvberg et al. (2003) who exposed how many of the participants in the process of procuring mega projects have incentives to underestimate costs, overestimate revenues, undervalue environmental impact and overvalue economic development effects.

Conversely, in an interesting experiment into communicating about risks to public health, Bennett (1999) revealed that there are a list of 'fright factors' which are most likely to amplify perceptions of risk associated with a new idea. These are when the risk:

- is involuntary rather than voluntary
- threatens a form of death or disease that conjures up dread
- damages identifiable victims
- is poorly understood by science
- is subject to contradictory statements from responsible sources of information
- is inescapable
- arises from an unfamiliar source
- results from manmade rather than natural causes
- causes hidden irreversible damage
- poses particular danger to children, pregnant women or future generations
- effect is inequitably distributed.

Finally, it is well known that in making judgements about the relative risks and opportunities associated with a new idea that people are influenced by the perceptions of other people around them. Teo's (2008) research on community opposition to construction projects showed how perceptions of risk associated with a building project are 'contagious' and can spread through groups and communities just like diseases. She also found that the process by which this occurs is largely invisible to outsiders and that once this process has started, it is extremely difficult to control. Ariely (2009) refers to this as the 'herding' effect – our tendency to assume something is better or more likely based on other people's behaviour and perceptions. For example, if we walk past a new restaurant which is full of people then we tend to assume it is good, despite the fact that we have no evidence to support that. It is also related to the Asch Effect (Asch 1951) where people's thoughts can become polarized in the direction of the shared values of a group or around the views of one individual dominant member. As Janis's (1988) research on 'groupthink' found, those who challenge dominant group norms are often classed as dissenters and rejected as an 'out-group'.

Risk management in complex systems

Throughout this book we have pointed to the increasingly complex nature of the innovation process. We now live in an intensely interconnected world with multiple and dynamic interdependencies and uncertainties which did not exist in the past. And as technologies continue to develop, it is likely that the pace of organizational complexity will accelerate at an exponential rate.

We have showed that fifth generation models of innovation have adapted to reflect these trends. However, traditional approaches to risk management have not yet adapted in the same way and are unable to provide managers with an adequate understanding of risk in such complex systems. Little has changed since Loosemore et al. (2005) argued, nearly a decade ago, that approaches to risk management produce a very static and linear view of risk which fail to explain how interdependencies within systems can amplify risk in unpredictable and often shocking ways. Recent global events have exposed yet again our inability to control the risks presented by the complex systems in which we are all now imbedded. The recent global financial crisis, triggered by the collapse of Lehman Brothers, is perhaps the most vivid example of how deregulated self-organizing complex systems are inevitably prone to sudden and catastrophic collapse. And natural disasters such as Hurricane Sandy and Hurricane Katrina show how catastrophic consequences can arise as failures in complex infrastructure systems unexpectedly cascade from one sub-system to the next. In both disasters, electricity outages caused telecommunications systems and internets to collapse, transportation systems to fail and critical services such as gas and water to stop, resulting in a multitude of unanticipated and uncontrollable knock-on effects to other seemingly unrelated systems and sub-systems.

The main difference between simple and complex systems is illustrated in Table 7.4

The main problem for managers in controlling the risks and opportunities in complex systems is their property of 'self-organization' – the ability of a system's connections and interdependencies to change, adapt and develop on their own without the influence of external managers. Complexity and chaos theorists argue that self-organizing systems tend to be highly efficient (compared to centrally managed systems) and find a natural equilibrium. However, in doing so they tend to settle at a 'critical edge' where a small change in the system can lead to catastrophic changes in the overall system. This is because in a truly complex system, each component will tend to act only locally, being unaware of its impact on the wider system. Koubatis and Schonberger (2005) vividly illustrates this with the analogy to a pile of sand. As each grain is added, it displaces the grains of sand around it but at some point, as more grains are added, one grain will act on the whole system and cause the pile to collapse and find a new equilibrium. In other words, complex systems evolve gradually and smoothly at a steady pace to be punctuated by occasional and inevitable catastrophic changes. While they may appear under control on the surface, in reality they exist on the edge of chaos and it is impossible to predict when the sudden change in equilibrium will come. These theories have been used to explain the behaviour of many self-organizing systems such as the stock market, traffic jams, epidemics, volcanoes, evolution and extinctions, forest fires and economic cycles. However, they have never been used to understand the innovation process.

Table 7.4 The characteristics of complex systems

Simple system	*Complicated system*	*Complex system*
Small numbers of parts	Large number of parts	Very large numbers of parts
Easy to understand	Difficult to understand	Uncontrollable dynamics
Linear connections which are controllable	Linear connections which are difficult to control	High degree of non-linearity
Static relationships	Relationships can change but predictably	Multiple interdependencies
Fixed parameters	Movable parameters	Parameters can change independently
Predictable outcomes	Unpredictable outcomes	All possible outcomes impossible to predict
Full understanding of how system works	Difficult to understand how system works	We do not understand how it works despite knowing all the parts
Easy to model	Difficult to model	Different models will produce different results

(Source: adapted from Koubatis and Schonberger 2005)

Using system dynamics to understand risk in complex innovation systems

While complexity and chaos theory can help us understand the risks of innovation in complex systems, the challenge is how to apply this knowledge in practice. It is easy for us to say that traditional approaches to risk management are inadequate, but what alternatives are available to innovators to better understand the risk they face?

One common approach is to adopt a stochastic approach using probabilistic models such as Monte Carlo simulation. In simple terms, this approach requires managers to list all the risks and opportunities associated with an innovation and to create a probability distribution for each risk and opportunity. The next step is to use computer software to randomly select points along each of those distributions for combining into one single cumulative distribution which represents the array of risks and opportunities associated with the whole innovation project. Modern computers allow this to be done rapidly over and over again, effectively simulating a project over many thousands of 'iterations' to provide a reliable indicator of the true probability of different innovation project outcomes. However, while this is an improvement on deterministic estimates of risk (which show just one possible outcome – the average outcome) the problem with this approach is that it still presents a static picture of risk at one point in time.

One approach which has the potential to overcome this limitation is system dynamics (SD), which has been used to optimize the delivery and design of complex projects and systems such as aircraft and critical infrastructure. The philosophy underlying this approach is to try to understand complex systems using simple models and the six broad steps involved are:

1. Describe the system
2. Create the model
3. Simulate and refine the model
4. Design alternative policies, strategies and structures
5. Educate and debate
6. Implement changes to policies, strategies and structures.

Step one – describe the system

Step one involves describing the innovation system, its component sub-systems and its interdependencies. This can be done using a Rich Picture Diagram (RPD) which in simple terms is a pictorial multi-layered representation of the real world using symbols to represent sub-systems and arrows to represent relationships between them. A typical rich picture diagram is depicted in Figure 7.1 and was recently produced to model the risks posed to Australian hospitals' infrastructure by extreme weather events. This RPD represents pictorially the various components of a hospital system which determine the effectiveness of response to such an event and shows that the system is far bigger than the hospital itself, extending deeply into the wider community. Any innovation introduced into this system to help alleviate this problem will therefore have to fit into this larger system and interact with other sub-systems such as aged care facilities, emergency services, telecommunications providers, suppliers of medical materials and equipment etc. In Figure 7.1 the nodes are simply pictorial representations of 'critical assets' which have been identified as risks to the hospitals being modelled with connecting lines representing dependencies (flows of 'resources') between them. For example, in Figure 7.1, the availability of key maintenance staff (a key risk) depends on the roads being open (another key risk) to get to hospital. The arrow represents the direction of interdependency and a flow of resources between these two critical asset risks.

In Figure 7.1 there are three types of critical assets:

1. Organizations – suppliers, external service providers, external authorities etc.
2. People – staff, patients, public etc.
3. Physical – buildings, plant, machinery, infrastructure etc.

and there are five types of resource dependencies between the critical assets:

1. Information – X needs information from Y to respond effectively.
2. Financial – X needs money from Y to respond effectively.
3. Power – X needs permission from Y to respond effectively.

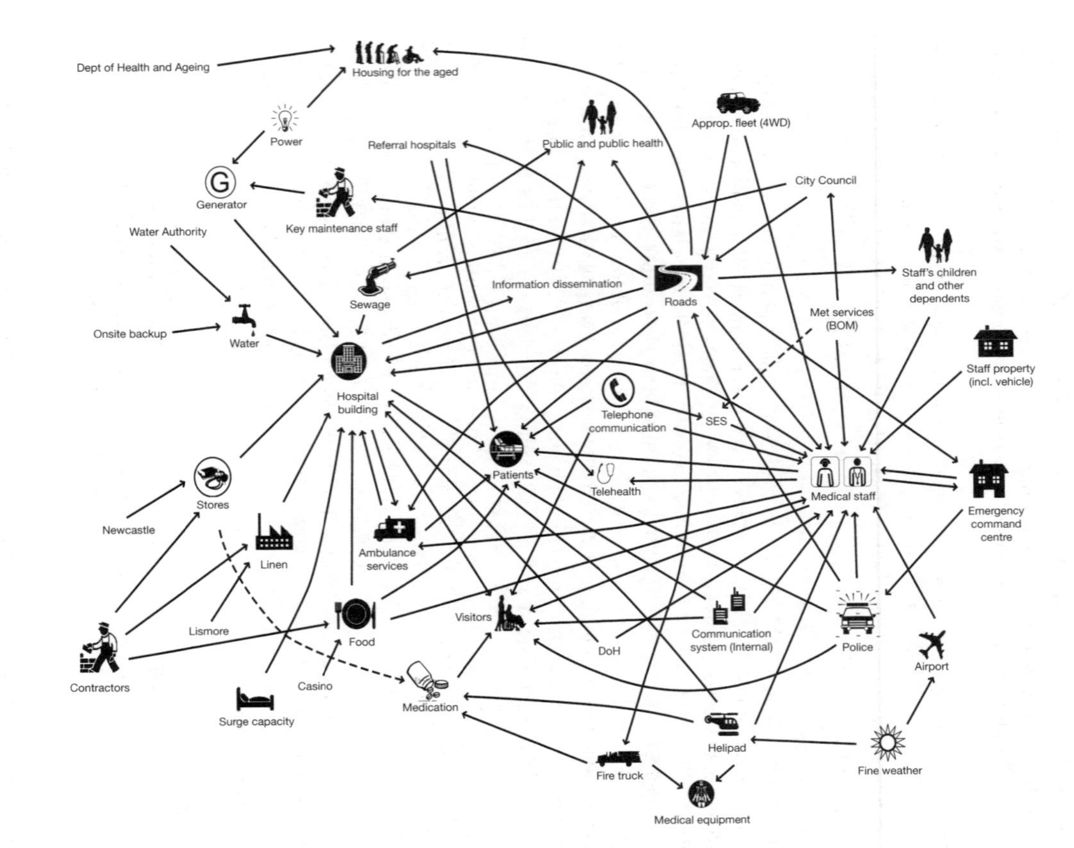

Figure 7.1 Rich picture diagram of a hospital system during and after an extreme weather event (source: Loosemore et al. 2012)

4. Material – X needs materials, water, energy etc. from Y to respond effectively.
5. Human – X needs staff, people, emotional support etc. from Y to respond effectively.

Step two – create the model

Step two involves formulation of a model to represent the system interdependencies and represent how the system works in the real world. The model maps the 'stocks' and 'flows' of key items of interest in the system. In simple terms, stocks represent accumulations of money, materials and information in the system and flows represent the rate of increase or decrease in those stocks over time as the system operates. For example, as patients in a hospital are admitted, treated and cured through health care services being delivered, the number of patients staying in the hospital rises and falls. A stock and flow model allows managers to experiment with the system structure to optimize its performance. By experimenting with different system configurations, managers can test the impact of different interventions on system performance over time. The stock and flow model developed from Figure 7.1 is illustrated in Figure 7.2 and was produced using a system dynamics software called Insight Maker. System dynamics software packages like Insight Maker provide a method for creating stock flow models which can be centrally located on a web server so that stakeholders can create the system design collaboratively.

Step three – simulate and refine the model

Step three involves running the model using system dynamics software such as Insight Maker and refining it by repeated returns to Steps 1 and 2 until the model becomes 'adequate' for the purpose under consideration. In systems theory 'adequacy' reflects the fact that there is no way to prove a model perfectly represents organizational behaviour in the real world. One can achieve only a 'degree of confidence' in a model balanced against the time and cost involved in further improvement. So it is normal to stop refinements of the model when there is consensus from stakeholders that the simulation represents how the system behaves in practice.

Step four – design alternative policies, strategies and structures

In step four managers identify alternative management policies, strategies and testing using the system dynamics model. Simulation tests are conducted to seek to determine which innovations show the greatest promise in achieving the objectives identified by the system stakeholders. The alternative policies, strategies and structures tested will rest on the imagination, experience and skill of the system stakeholders generating powerful alternative solutions.

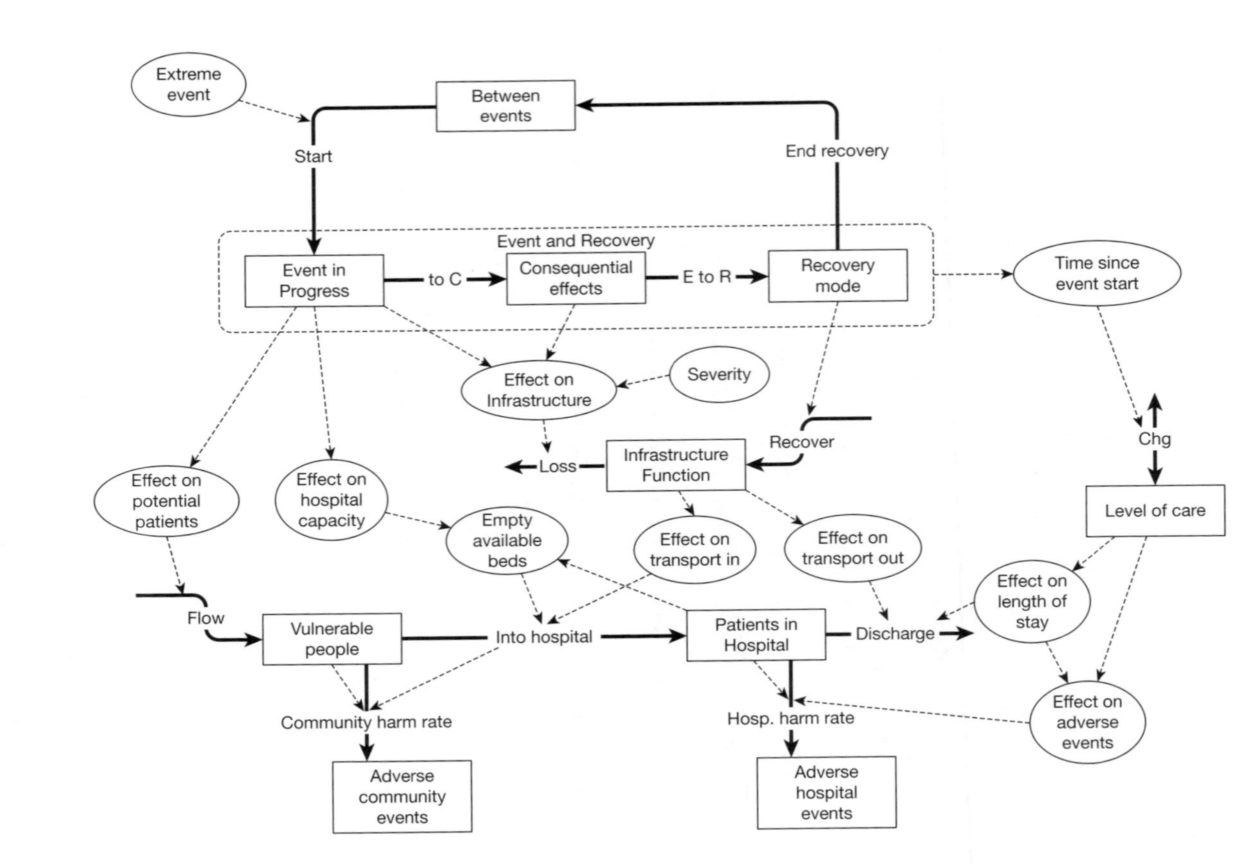

Figure 7.2 Stock and flow diagram (created using Insight Maker)

Step five – educate and debate

The simulation model will show which innovations are likely to produce the best combined outcomes for the stakeholders involved. As we have shown, the implementation of any innovation, based on this simulation, will normally involve reversing deeply embedded policies and systems, replacing existing technologies and physical infrastructure and changing strongly held emotional beliefs. Despite the fact that stakeholders have agreed that the system dynamics model accurately represents their system, there may still be reluctance to change. To overcome this resistance requires sufficient duration and intensity of education and debate to reverse traditional practices. Questions may also arise that require repeated recycling through steps one to five and no matter how many people have participated in these steps, it is likely that new stakeholders will need to become involved in discussions about implementation.

Step six – implement changes in policies, strategies and structures

If steps one to five have been undertaken successfully, there should be minimal resistance to implementation. Even so, implementation may take a very long time. Old strategies must be changed, old technologies replaced, people must be trained to work in different ways and new systems must be created and introduced to support the new ways of working. Evaluation of the changes should also take place during and after implementation to enable ongoing refinement of the simulation model to better reflect reality.

Conclusion

In this chapter we have tried to communicate the complexity of managing innovation risk in a world which is becoming increasingly complex. We have discussed some of the many risks which firms need to manage in making their innovations a success and we have argued that the only way to understand these risks is to adopt a holistic perspective. Yet traditional approaches to risk management take a narrow view and teach us that best way to understand risk is to 'narrow it down' and to take a specific view. It is not surprising therefore that so many innovations fail. The problem for managers in overcoming the limitations of traditional approaches to innovation is that there are few tools available to model the interdependencies in complex dynamic systems. However, in this chapter we have shown that one approach which can help innovators understand the big picture of the intertwined risks with an innovation is system dynamics. The system dynamics approach is a potentially powerful way for organizations to work collaboratively with its stakeholders to collectively optimize system configurations to ensure that an innovation has the maximum chance of success.

8 Conclusion – we need a reality-check

If you want something new, you have to stop doing something old

Peter F. Drucker

Bringing it all together

We conclude this book by bringing everything together into a model which managers and policy-makers can use to simplify and manage the innovation process in their businesses or countries. The model in Figure 8.1 is a summary of the latest research on innovation within and outside the construction industry. However, like every model, it has its limitations and to pretend that the real world operates exactly in this way would be naive to say the least. So to reality-check the model, the practicalities of implementing these ideas in practice are tested against the views of leading innovators in industry and government to conclude with a model which can be relied upon in practice. Figure 8.1 should be read from the bottom-up and the key elements of the model are summarized below.

Encourage it

During this phase, managers create the enabling conditions which allow innovation to occur. A business culture is created where people feel they can innovate and where they instinctively do so without consciously thinking about it. There is a demand and supply side to this process. The supply side is the responsibility of business and the demand side is the responsibility of clients and government regulators.

The supply side

Strategize:

- Link innovation to strategic corporate objectives (it should be clear how innovation contributes to core business goals and corporate identity).

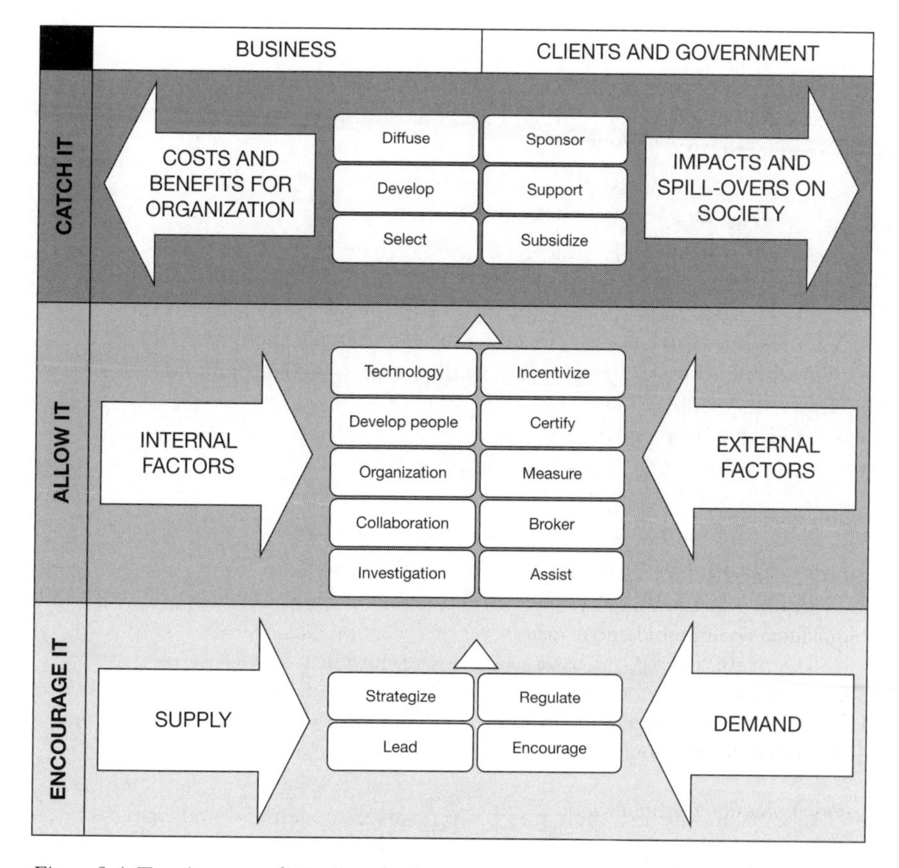

Figure 8.1 Turning serendipity into business success

- Have a strategic focus on innovation (make innovation a priority, articulate a vision about innovation, decide areas where innovation will happen, communicate clear and measurable innovation goals, have a clear plan to achieve them, decide how to measure innovation success).
- Focus innovation strategy around the desire to solve a problem, to satisfy a need or to address a specific cause. Balance economic, social and ecological goals.
- Develop and balance resources across a diverse innovation portfolio which covers a range of ideas developing in parallel and which addresses different market segments.
- An innovation strategy should be adaptable and present a set of flexible strategic choices rather than a rigid set of goals. People need space and discretion to innovate.
- Innovation strategy should be built 'bottom-up' rather than imposed 'top-down' (stakeholders should be involved).

Lead:

- Build a culture of tolerance, transparency, trust and openness (which enables people to put forward ideas in confidence and to take calculated risks without fear of blame or failure).
- Use corporate identity as a unique resource to inspire difference.
- Inspire (create a vision that 'attracts' new ideas).
- Question (challenge existing and accepted ways of doing things).
- Listen (new ideas should be able to come from anywhere at any time).
- Discriminate (able to recognize the good ideas from the bad).
- Consult (involve people, listen to them and make them feel valued).
- Integrate (break down business silos and ownership of knowledge by specific groups).
- Aspire (encourage people to push boundaries and explore new paths).
- Forward looking (always looking for opportunities in over-the-horizon trends).
- Determined (overcoming barriers and resistance to change).
- Driven (by a strong cause and set of values and beliefs).
- Flexible (in response to unexpected opportunities and constructive feedback).

The demand side

Encourage:

- Clients should use their purchasing power to demand and expect innovation and set demanding goals which stretch firms and cause them to think in new ways.
- Clients should create a competitive environment which recognizes and rewards innovation (they should employ on the basis of best value not lowest price).
- Clients should treat innovators and their ideas with integrity and respect.
- Governments should seek to reduce the risk of innovation by providing a stable and conducive economic, legal and regulatory environment for innovation to occur.
- Governments can stimulate innovation through competition and innovation policies and through their education and immigration policies which provide a talented pool of human resources to provide innovative ideas.
- Governments are the only actor capable of taking an overall view of a national innovation system and need to develop national innovation policies and strategies to guide the private sector and connect the various elements needed to make it work.
- Governments should design and coordinate their approaches internationally to find global solutions to trans-national challenges. Many of the problems we face today (like climate change) are so complex that they can only be

solved at an international level with international governmental coordination and collaboration.

Regulate:

- The construction industry has a long tail of under-performing firms which will not innovate unless forced to do so. Therefore government regulation is essential to drive change and encourage innovation around it.
- Regulation also works by making it difficult for poor performers who undermine innovation to operate. By cutting-off the long tail of under-performers who drag standards down, innovation is more likely.

Allow it

Having created an environment where people feel comfortable to innovate and a duty to do so, the 'allow-it' phase is where new ideas germinate and are developed. The quality and quantity of ideas produced and the speed at which they develop depends on a range of internal and external factors. The internal factors are organizational whereas the external factors are determined by the business environment.

Internal (organizational) factors:

Investigate:

- Free-up time for people to think (build-in redundancy).
- Encourage imagination, non-conformity, dissent and creative destruction.
- Develop a peripheral vision which recognizes the value of information from diverse sources (creativity best happens at the intersection of different knowledge domains).
- Encourage curiosity and a thirst for knowledge (applied and pure).
- Invest in research (to look over-the-horizon, to explore new ideas, to generate new knowledge and to dig for unmet needs).
- Develop an ability to learn new practices, adapt existing ones and unlearn old ones.

Collaborate:

- Emphasize collaboration rather than competition. Get people involved early in the development of solutions.
- Build a dense 'spider web' network of relationships that fosters a diversity of connections and 'catches' lot of different pieces of information from different people with different perspectives.
- Share resources and collaborate actively in strategic alliances around innovations (with firms inside and outside the industry).

- Recognize the value of weak ties (it is often with those with whom you least interact where your greatest value might lie).
- Build social capital by developing unique inter-firm networks with other innovators (in supply and demand chains).
- Share risk and reward through integrated supply chains (which allow innovators to benefit from their ideas and move them closer to customers and users).
- Build connections across traditional silos and knowledge domains.
- Build organizations with permeable boundaries 'at all levels' to maximize information exchange with the business environment (two-way communication).

Organize:

- Develop systems to support innovation.
- Build loosely coupled, decentralized organizational structures rather than centralized mechanistic structures.
- Build integrated teams which stay together for the duration of a project.
- Challenge traditional professional boundaries, roles, responsibilities and knowledge silos.
- Organize around opportunities.
- Organize across boundaries by avoiding rigid organizational silos which act in competition with each other.
- Develop agility (an ability to adapt in response to unexpected opportunities).

Develop people:

- Deepen and widen the gene pool. Build depth (through talented people) and breadth (through diversity) of talent.
- Develop people – give them the skills and knowledge to innovate.
- Recruit, empower, incentivize and reward innovative people.
- Recognize that a range of skills and attributes are needed at different stages of the innovation process.
- Use flexible job descriptions which enable people to adapt their jobs to changing circumstances and opportunities.
- Provide slack-resources, time and spare capacity. People need to adapt and think laterally.
- Build engagement – an employee's sense of identity with an organization, their positive or negative emotional attachment to their job and to their colleagues profoundly influences their willingness to learn and perform at work.

Technology:

- Explore new technology (as a means to innovation not an end in itself).

- Know 'why' technology is being used, 'how' to use it to competitive advantage and 'when' to incorporate and integrate it into business processes.

External (environmental) factors

Incentivize:

- Create incentives by rewarding firms which innovate.

Certify:

- Establish certification schemes to label innovation, set minimum standards and assure customers.

Measure:

- Set clear and ambitious stretch targets to encourage continuous improvement and establish common metrics to allow firms to measure innovation performance using standard comparable methodologies.
- Measure, monitor and benchmark innovation performance.

Broker:

- Build social capital by setting up inter-firm knowledge sharing networks, connecting innovators with venture capitalists and customers, organizing trade missions etc.

Assist:

- Reduce red tape and simplify procedures around innovation.
- Fund research.
- Provide R&D tax offsets and allowances for innovative firms.

Catch it

Having created the environment for innovation to thrive and having generated many new ideas, the 'catch-it' phase involves spotting a potentially marketable idea, developing it into a business opportunity and implementing it effectively into the market to produce the returns anticipated. This stage is all about timing and understanding when ideas match market needs. This doesn't happen in a linear predictable way. Ideas can get 'stuck' in the allow-it phase for many years without becoming a business opportunity. Indeed, some ideas may never be implemented and the effectiveness of this process depends on the following.

Selection:

- Understanding market needs (through market intelligence and good customer relations and feedback).
- Spotting ideas which can match these needs (have market resonance).

Development:

- Converting ideas into buyable products and services which match market needs (through financing, design, adaptation, prototyping, testing and refinement).
- Develop business models around new ideas (sales volumes, margins, revenues, markets etc.).

Diffusion:

- Marketing – understand what motivates customers to engage with an innovation, know how to communicate innovations to markets where their value will be most appreciated.
- Selling – know how to get customers to invest in the innovation (use opinion leaders and change agents etc.).
- Delivery – know how to achieve saturation in the market (through widespread acceptance before competitors copy and enter that market with similar ideas).
- Servicing – delivering on on-going support.
- Learning – collection of customer feedback and adaptation of new ideas to stay ahead of competitors.

Externally clients and governments can:

Sponsor:

- Act as champion, provide patronage, act as guarantor, provide support etc.

Support:

- Provide market intelligence, marketing advice, networking, brokering connections between firms and customers etc.

Subsidize:

- Provide finance, seed funding, venture capital, underwriting etc.

Reality-checking the model

The model in Figure 8.1 is a summary of the latest research on innovation within and outside the construction industry. Like every model, it has its limitations and to pretend that the real world operates exactly in this way would be naive to say the least. So to explore the realities of implementing these ideas in practice, this section solicits the views of senior executives and government advisors who have been responsible for driving innovation across the construction industry and other industries around the world. These people were presented with the model and asked to comment on its relevance in practice. Given the diversity of views received, it is impossible to capture them all here. So only those which represented the views of the majority have been presented. Rather than associating specific quotes with individuals, these discussions are presented in the author's own words. However, where a particular comment seemed to be relevant, it has been included to enliven the text without attribution.

Encourage it

There was agreement that an organization has to be 'set up' for innovation but divergent opinions of how this is best done.

Leadership and strategy

While there was a sense that strategy and leadership were important, there was also a feeling that innovation is often portrayed rather romantically through 'rose painted glasses' and that much of the literature in innovation is post rationalized: 'everyone needs to take an honestly pill', 'Entrepreneurs are egotistical and will tell you about the culture they created after the event. It will always be a reflection of their personality . . .'. Also, stress, hardship and challenge are key ingredients of innovation. Much innovation starts with discomfort – 'you can't comfortably innovate'. If managers are happy with their situation then there is not the pressure to try something new. Where there is most discomfort most innovation will tend to occur. If someone has seen hardship then they are more likely to have the toughness, self-belief and determination to succeed in the face of the inevitable resistance that will come from innovating. Innovators are not just smart, they are 'hard nuts, egotistical, arrogant and determined'. However, this must be balanced by compassion: 'people don't follow aggressive leaders. If you care about them they will follow you'.

Having said the above, there was widespread agreement that innovation needs to be defined, the reasons need to be identified and benefits need to be spelt out for everyone involved. This was seen as the primary role of strategy – to create a 'pre-disposition to innovation', to formalize the process in some way, to place it at the centre of business priorities and resource it properly. And there was agreement that this must start at the top, rather than be bottom-up. The bottom-up element of strategy comes from the importance of not having a rigid plan but a high degree of flexibility to provide people in lower reaches with room to move within the wider goals set by leaders: 'you say what you want to be and what you want to achieve and then let people figure out how to get there', 'you can't innovate by going on course innovation 101 . . . it's all about people seeing and acting on opportunities', 'Entrepreneurs don't run a process, they just do it . . .They do what needs to be done'.

However, while innovation is a core value in the corporate strategies and public statements of most firms these days, there were divergent views on how important it was in leading to innovation outcomes. For example, a formal strategy was seen to be more important in larger firms than smaller firms. Furthermore, while some pointed out that it was key to 'giving people permission to innovate' and communicating that 'innovation matters', others qualified this by simply stating that 'we will innovate' would be unlikely to produce any up-turn in innovative capability: 'statements don't mean anything . . . they must challenge people to innovate by setting challenging targets and innovative companies constantly challenge these statements and look within to ask where they will do so'.

In general, there was consensus that innovation won't happen unless the leadership is there to realize that vision by enacting the strategy: 'People need to see the organization does new things – visibility is critical . . . you can't tell people . . . you need to show them'. And it helps if the leader has a proven track record: 'if you don't have runs on-the-board, why would they follow you?'. Furthermore, since innovations often challenge existing power structures and 'steps on someone's toes', leadership is considered essential to impose the authority to overcome that resistance.

Most of the respondents also thought that the importance of being seen to be innovative in today's business environment meant that having an innovation strategy too often became an end in itself, rather than a means to an end. To ensure this does not happen, the job of the leader is to clearly articulate the reason for innovating: 'having a project to win, a problem to solve and a cause which people can rally around is important . . . Innovation won't happen if someone simply dictates that it will happen'. Having a focus and a reason to innovate provides the thirst for knowledge that drives innovation: 'it opens your eyes and encourages you to go looking for new ideas'. Strong, focused and determined leadership is critical in overcoming many of the inevitable barriers which will confront innovators in the construction industry. Without leadership 'innovation doesn't get an agenda or quantum' and 'is directionless'.

There also appears to be a definitional problem which leaders need to overcome. To most people innovation is a rather nebulous term: 'most people when asked wouldn't know what innovation is'. So it was felt that leaders need to give innovation a focus 'to make it real'. Importantly, the extent to which innovation is internalized by staff is also determined by the 'acting out' of those values by senior managers. Leadership is essential to 'give life to those values' and they also need to be articulated in measurable terms so that they 'mean something' to people.

As well as providing focus, most thought that the role of the leader was to build an innovation culture: 'they set the tone in a company and the risks associated with failure'. An innovation culture was seen as critical by everyone. This should not only define a particular way of thinking but it should also ensure that there is a level of maturity and continuity to the process. This comes down to five main things:

1. A clear vision which makes innovation important
2. Leaders 'walking the talk' (determined, inspirational, energetic, visionary, open)
3. Building human capacity to innovate and supporting innovation capability
4. Rewards and support for innovation initiatives
5. Creating and nurturing a sense of community and collective identity which people can rally around.

Corporate identity was considered to be especially important in creating a sense of collective responsibility for innovation. Many firms could trace their corporate identity back to a founder who, in building up their companies, established a particular set of values and way of thinking which persisted to this day. The fact that these very different corporate identities exist indicates that there is no one innovation recipe and that every firm needs to find and cultivate its own unique innovation focus.

Integration

While the concept of integration was widely considered as critical for innovation, many agreed that the ideal of integration is often unachievable in practice and that it is important not to lose sight of the need for business accountability. As one of the respondents said,

> One of the big mistakes people make in pursuing integration is to assume that one has to abandon accountability in the pursuit of trusting and open relationships. There is a mistaken assumption that being 'business-like' destroys trust whereas in reality the opposite is true. While integrated teams are important to drive innovation, if done without regard for proper accountability then it can often be a 'bridge too far' for clients.

One of the suggested reasons for this is that many directors on the boards of clients are lawyers, which creates a powerful reluctance at governance level to move away from traditional approaches which appear, at least in the short term, to offer some form of accountability and control. It is critical that in pursuing innovation and interpreting much of the literature in this area, ideas from other sectors must often be tempered by the reality of the construction industry. The imposition of a completely open system onto an industry that is 'fundamentally dysfunctional' could potentially create more problems than it would solve.

Planning for innovation

Only one respondent had seen an innovation plan and only one had an innovation unit. In most cases, innovation occurred out of necessity – either of winning a job by adding extra value or in solving an immediate problem on a project. It wasn't a highly planned or formalized process. One of the biggest challenges in managing innovation in construction is that there are permanent business organizations managing temporary project organizations. As one respondent said, 'the top business layer is relatively stable and there is little innovation that occurs here but underneath it is a fluid project-focused undercurrent which is highly dynamic, subject to a lot of uncertainty and unexpected problems which have to be responded to'. Both layers require very different approaches to innovation. The top layer will tend to have a longer time horizon and focus on innovations which span different project boundaries. In contrast, the project layer will tend to be very short term and client focused. People operating at this level tend to focus on innovations which can deliver productivity and cost benefits to deliver the project on time and on budget: 'most companies focus on the incremental operational innovations and the big game-changing innovations that provide competitive advantage are few and far between . . . it's hard to think of them in construction'.

The role of clients

'Clients play a critical role in the innovation process since they pay for it' and innovative firms strive to 'attract themselves to clients that value innovation'. However, it was widely felt that most construction clients are not open to innovation or prepared to pay for it: 'most clients are completely irrelevant to innovation. They have no interest in it whatsoever. Unless of course it can reduce costs . . . then they have a great desire for innovation'.

Many felt that the education of clients is an important part of the early innovation process. Unfortunately, it seems that many clients are difficult to educate and think too short term to allow innovation to occur. There was a sense that they typically prefer to minimize short-term capital costs, go to tender and maximize predictability of outcome by transferring risks to the construction industry. As a number of respondents complained, 'by the time we are involved in a project it's too late . . . the solution is locked in and the value-adding opportunities are reduced to a trickle', 'we have a fundamentally dysfunctional industry which is the epitome of destructive competition. It is highly fragmented and has refined itself down to the point where innovation can't happen because there are no resources to do so'.

The role of government

There was universal agreement that the primary reward for innovation must come from the market not from government incentives. Clients are critical because if there is no one to sell to, then most people won't innovate: 'it's hard to shift or create whole markets . . . it's much easier to respond to a need', 'if they don't want your idea then you haven't got a value proposition . . .', 'if clients do not want energy efficient buildings then the industry won't build them. Similarly, if a firm isn't focused on innovation and set-up to innovate then incentives will have little impact in encouraging them to do so'. Insensitivity to the realities and needs of markets was seen as a major reason why many innovators fail: 'this is why many backyard innovators fail . . . they invent in a bubble'.

Having emphasized the role of the market in driving innovation, good government innovation policy was widely seen as critical to 'stimulating the conversation' and 'setting the enabling environment in which innovation can occur', 'where firms can profit from innovation' and markets can operate effectively. Governments were seen to play a critical role in providing innovation infrastructure, capital (financial, physical and social) and regulation to encourage change and to protect innovators from 'the race to the bottom'. Incentives to innovate were mainly seen as being market driven rather than government driven. If the government introduces greater tax concessions for R&D then firms will tend to respond, but the impact is relatively small. For example, definitions of R&D are often so ambiguous that many firms simply re-badge existing initiatives as innovations to secure the largest tax rebate for the minimum investment of resources: 'Many firms simply see R&D tax concessions as another opportunity for creative accounting which allows them to squeeze more money out of their operations rather than to innovate'. The government can also set minimum standards through certification systems and codes of practice, 'although they can also ossify and inhibit innovation by constraining thinking if badly designed and managed'. To avoid this, certification systems must not be prescriptive but must be performance-based and give people space to innovate in achieving those performance standards.

While government incentives and regulations are useful to 'underpin' innovation, the main role of government is at a macro and systemic level in creating a national agenda around innovation and innovation policy frameworks to support innovation. There is also an important capacity building and brokerage role for government which is the current domain of policy action at the moment in helping to up-skill and connect firms to other complementary organizations and businesses in domestic and international markets. This is particularly important for the many small-to-medium-sized enterprises which dominate the construction sector, which are often lacking in management expertise and the experience and confidence to connect to other leading firms which may help them to grow their business. The Australian Government's Enterprise Connect initiative is a good example of this type of assistance. Enterprise Connect offers business mentoring, facilitation, advice and support to Australian small and medium businesses to help them transform and reach their full potential. This includes free business reviews, grant assistance to implement recommendations and a range of tailored innovation services to meet individual business needs.

Through the Enterprise Connect Researchers in Business initiative, the Australian Government also supports the placement of researchers from universities or public research agencies into firms that wish to develop a new idea with commercial potential. Finally, in addition to being a 'supplier of innovation capability and risk sharer', governments also play an important 'demander' role as major construction clients in their own right. However, while governments should aspire to be a 'best practice client and employ on value not lowest price', this is often difficult in practice: 'it is hard for public clients to reject the lowest price since they are accountable to the taxpayer' and 'clients generally lack the methodologies to value the innovation in a non-conforming or innovative bid'. Furthermore, 'clients will invariably opt for some certainty up-front which means that by the time we get the chance to get involved, design solutions have been set and the decision to opt for an alternative solution will often result in an unacceptable loss of face and waste of resources'.

Governance

'Governance is too often forgotten in the innovation discussion.' Governance was thought to influence innovation in five main ways. First, board culture influences business culture: 'If the board is open to new ideas then this will filter down through the business'. Second, it is important to have properly structured boards with the right mix of experience, insight and expertise: 'we have deliberately created a diverse board to ask us difficult questions that cause us to think in new ways'. Third, effective boards should be highly accessible and act in a strategic advisory role to business units with good ideas: 'we can ask our board questions any time. They act as a sounding-board for us'. Fourth, for innovation to flourish, boards must be strategic and not just focus on financials:

> financials create a very narrow mindset but it's hard for large organizations to get away from them and think more broadly. The idea that one has to grow continuously year-on-year can be very destructive. Why can't firms focus on becoming more efficient year-on-year?

Finally, boards have a responsibility to ensure that there are adequate risk management capabilities and systems in place to allow people to make decisions with the confidence that the risks involved will be managed effectively: 'we have good systems in place to manage our risks. This allows us to experiment with new ideas with confidence'.

Allow it

People are critical

There was agreement that innovation revolves around people. However, there is a need for different types of people: 'at the start there needs to be a predisposition to innovation . . . then there is a need to hand over to people to make them a reality', 'there is a need for people who can think outside the square and people who think inside the square'. As one respondent said,

> We are more-or-less the same as every other competitor apart from our people . . . They define our identity and we structure our business around them. If we did it the other way around we would be forcing our strategy on our people and it wouldn't make any sense to them.

While an innovation strategy must have a structure and focus, most felt that it was also important to ensure that it is broad enough to allow people to move within it. Innovative firms tend to have a loose and mobile business strategy which is built around their people's capabilities, desires and passions. As their people develop so does their strategy and this enables firms to reinvent themselves and gravitate to where the work is in an ever-changing market. In most instances, this is not a structured process: 'It just happens'. This means that it is not uncommon for new business units to emerge out of the personal interests and passions of individuals: 'you find out what people are interested in and help them explore those ideas'. For example, in one firm, the passion of one staff member for horses led to a successful business unit which specialized in equestrian stadia. In another firm, a new digital innovation business unit was born out of a PhD completed by one member of staff.

Most of the firms have a core team of people who lie at the heart of their business and who have been there for many years. These people form a tight network and they define, protect and drive the business culture:

> there is a kind of nest . . . a central nest which has a deep thirst for knowledge driven by the high levels of competition in the industry and the continual need to create a competitive edge. The nest is a highly competitive environment in itself and one has to intimately understand the business culture and demonstrate commitment to it to be accepted into it . . . once you are in the nest you become part of the family and they feed off each other so it's hard to identify an individual leader. The next layer out moves between different firms in the industry and they are all trying to become part of a nest. It's the central nest that inspires difference not the outer layers which are far more mobile and less attached to any individual corporate identity.

The forgotten layer of people in the industry who some felt were largely ignored in the innovation process were the construction workers who undertake the work on site. It was felt that the drive for greater flexibility in the industry over many years has resulted in them being forced into casualized labour, reducing their ability to participate in workplace issues: 'The structure of employment in the construction industry is all about lowering costs and providing flexibility rather than being about innovation. Contrast this with the manufacturing sector where in a factory of one hundred workers there are all sorts of opportunities for them to contribute ideas and to talk to managers', 'it's a race to the bottom'. Many respondents pointed out that one of the major negative consequences of this employment structure is that the subcontractors who are expected by the large builders to innovate are largely separated from the benefits of doing so by repeated subcontracting and by a widespread 'risk transfer mentality' which ensures that those who need to innovate are rarely able to benefit from them. Many also pointed to the lack of investment training in the industry which has resulted in a stagnation of ideas. Firms which provide apprenticeships and develop their staff are in the minority and can place themselves at a major cost disadvantage to those which do not. This has had a huge impact on innovation by effectively freezing the intellectual development of the industry. As one respondent said, 'The basic skills levels of workers in other industries are far higher generally than they are in construction. The knowledge base of workers in construction has hardly developed in the last twenty years because of the subcontracting revolution'. This creates a huge disconnect between the ambitious and imaginative ideas which are concocted by large contractors and consultants and the ability of subcontractors to deliver them. The subcontracting model appears to have created a two-tier industry of 'haves and have-nots'. While tier one subcontractors are highly professional and competent, there is a 'skills-cliff' when one enters into the second and third tiers where standards of leadership and management fall-off very quickly. One respondent pointed out that

> A lot of the small to medium sized business in construction industry are owned and managed by people with no formal education. They normally start as a trade's person and then become a manager by default as their business grows. Most have a trade qualification but few have acquired any formal management education. So while the multinational firms in the industry are as well managed as in any other industry, their ability to innovate and reform the industry is limited by a very long tail of poorly managed small firms which generally undermine innovation and performance . . . although there are some very good trades people in the industry at subcontractor level, there are fewer good managers and businessmen. So they just keep doing the same thing. They are locked into the past and are unlikely to embrace or even know about new technologies and ideas which could transform their business.

When the market is tight the skills-cliff within the supply chain is not a problem because there are enough tier-one subcontractors to go around, but when times are good, contractors are forced to draw on these outer layers to deliver the increased volume of work. However, there is always the temptation to draw on these outer layers since these firms can often offer a lower price and allow contractors to win a project. One of the respondents described this as 'a race to the bottom'. This explains that while firms have been refining their performance and getting more efficient, improvements have been very incremental and there have been precious few major innovations delivered by subcontractors in the last two decades. To address this problem one contractor was working on education initiatives which would help up-skill its tier-2 and tier-3 subcontractors. However, it also knew that without a special relationship with those subcontractors, its direct competitors will benefit equally from its investment. A collective cross-industry solution to the problem is therefore essential.

Up-front and down-stream innovation

Ideas are generated from two main sources: up-front competition to win jobs and the down-stream need to deal with a problem on a project and deliver it safely, on time and within budget. As one respondent said, 'Up front innovation is about winning the job and is driven by the need to beat the competition. But once you have the job it switches to how to deliver the project faster and more efficiently'. Up-front innovation is the type of innovation most discussed in the innovation literature since it involves looking through various solutions to best meet the client's needs. However, as a project moves towards implementation on site, it is clear that the space for innovation is reduced as the focus switches to delivery. As one respondent said,

> Innovation at a project level must be practical . . . You can't have too much creativity on site since the concrete has to be poured. There are basic things that need to be done and time is so tight that there is little time for creativity in doing things differently . . . where the innovation happens on site is in dealing with problems to keep the programme and budget on target. This is a different type of innovation – it is reactive not proactive.

So it is clear that different types of innovation occur at different stages of the project life-cycle and require different management strategies. But it is also clear that the best project managers can overcome the constraints on their projects even in the most testing of project environments. As one respondent said, 'No matter how pressured the project, good project managers always seem to have time to tweak things. They are highly disciplined entrepreneurs who can identify a good idea early on that holds value for their client and understand its implications for project delivery'. What differentiates these project managers is their understanding of customer needs and the quality of their relationships with customers who trust them with new ideas that may otherwise be rejected.

Idea generation

As one respondent said, 'Innovation is a nice word . . . it's all fine to say that managers need to build-in redundancy and give people time to read and explore ideas but you need a highly sophisticated management team to do this'. Furthermore, the idea of creating defined slots of time to search for ideas separates it from the day-to-day activities: 'you can't separate innovation into distinct time slots and expect people to innovate between one and three . . . it must be an integral part of their day-to-day jobs', 'innovation is not a discretionary activity', 'one often reads about people sitting around on bean bags etc. I wonder how long these models survive when a company hits some financial stress'. One respondent pointed out that their innovation unit had been ring-fenced for a period from wider cuts in the business because the owner believed it was essential to business success. But they also conceded that their success was hard to measure and that they had to report and justify their existence.

There was broad agreement that diversity is good for innovation: 'it provides a variety of perspective to consider a problem'. However, there was also a view that diversity was 'the luxury of large organizations' and that the best way for smaller firms to foster diversity was 'to replicate this through collaboration' with other organizations with complementary skills. There was also agreement that innovative firms seem to constantly fill their time in searching for new knowledge in as many forums as possible. One respondent likened these firms to 'mini universities'. They develop and nurture their knowledge networks and there is a culture of talking and debate. People also tend to wear many hats and constantly search for new ideas which may inform new ways of doing things for themselves but also for their colleagues. Typically, in these firms there are lots of new ideas being explored at any point in time and lots of work is done in people's own time. People are dedicated to the business and are continually thinking about new ways to improve it. However, it was also acknowledged that people didn't read enough and that it was very hard to find time to do so: 'we use the downtime between jobs to do this'.

Ideas are generated via a wide range of mechanisms. For example, one company had established a dedicated innovation unit to essentially act as the brains of the company. Another company provided one day a month where staff could work on new ideas. The best ideas are then supported by an internal research grant system. Numerous companies make new ideas accessible to everyone in the business through intranets or global crowd sourcing platforms. These tools put new ideas into a pool where people can read about them, contribute to them and even vote on their business potential. Other common idea generation mechanisms include structured project-specific discussion forums and lunch-time events where staff can listen to outsiders – often from completely different industries. These often take the form of formal Q&A forums led by international experts and senior managers which are podcast to staff. One company had a three-phone call rule whereby everyone should be able to get help in no more than three phone calls and a communication network which allowed people to ask questions across its international operations. A number of firms have created an executive position specifically to drive innovation. As one senior executive pointed out: 'This person's job is to look at the business again and all its processes . . . We took him out of an operational role where he didn't have to worry about delivering projects. The idea was to create a slack resource who has the time to think and connect the dots and capture knowledge across the business'.

While some companies have a highly structured approach to generating and selecting new ideas, in others this occurs by osmosis. For example, in one company, ideas develop organically and percolate slowly to the surface through 'socialization' until they become part of the accepted vocabulary. This process of 'natural selection' allows ideas to be scrutinized at multiple levels enabling ideas to be 'tweaked' and everyone to build up a stake in them. Through socialization new ideas move from being an individual construct to becoming accepted as the language of the wider organization. Only when an idea has passed this scrutiny is a business case developed. While this may be slower and less prolific than the structured approach, the advantage of the socializing approach is that it increases the chances of success for ideas that do manage to rise to the surface.

Collaboration and networking

Collaboration and networking were seen as quite distinct activities. Networking was seen as something all good innovators do: 'Entrepreneurs are a call away from the help that they need', 'good networks are very important to minimize the very high transaction costs associated with innovation'. Entrepreneurs are seen as highly 'active' people. They build, nurture and use their networks to advantage: 'they can make a phone call and get someone to help them, rather than paying (a top consultant firm) . . . their networks are gold. They are very important'.

Collaboration is a more formal process which was widely seen as crucial to innovation: 'it's very rare for successful innovators to work in a vacuum', 'collaboration is critical to get ideas across the line. The more people you can take the better'. However, when talking about collaboration, there was also some cynicism.

> Collaboration is just a word . . . there is nothing new in this . . . what it really means is that innovation comes from activity . . . activity spurs mutation . . . if there are more people working together on something then there is a greater chance of new ideas happening.

In construction, competition was widely seen as a barrier to collaboration: 'you can lose that competitive edge if you involve subcontractors too early. It can cause you to lose the job'. Competition is not only a problem for subcontractors. It is also a problem at a higher level: 'collaboration in construction is a huge challenge. There is always a sense that one is giving away something. The industry is so competitive that collaboration between the few big firms is almost impossible'.

For true collaboration to happen in construction, most argued that it was important to be involved early in the design process where solutions can be jointly developed. Most of the respondents talked about 'going on a journey', 'early-on the door is fully open to innovation . . . but this is a highly protected place to be and most people are faced with a firmly closed door'. As one respondent said,

> on one recent job we saved the client $20 million by redesigning the building. The client was open to it and willing to throw away the original design. It was a big thing to do. We demonstrated it again on another tender and it didn't work. The client had invested too much time and there was a big loss of face in acknowledging that the designs could be much better.
>
> More than anything else, clients appear to want certainty of price, there is a mistaken tendency to think that this is best achieved by staging the building process into distinct stages and passing all their risk down the contractual chain which is not conducive to collaboration . . . Many client advisors lack the courage and imagination to think outside the box . . . they are only interested in short-term goals . . . there are multiple layers of distrust within the construction industry.

So as a number of people said, 'if your innovations lead to productivity and cost savings then clients will pay'. Of course, in reality they are not paying at all.

Another issue raised by respondents was the lack of depth in the subcontracting sector which makes collaboration difficult to implement in practice. The subcontracting market is dominated by a relatively few large firms and there is a long tail of tier-two and tier-three firms which have very low management capability. Collaboration can therefore become highly incestuous as competitors draw on the same pool of subcontractors to do their work. Innovation in this environment comes out of the 'quality' of the relationship formed with a subcontractor. However, this is also problematic because no subcontractor is able to commercially survive if they tie themselves into a special relationship with one contractor. On the other hand, no contractor wants to get too close to a subcontractor because they lose the element of competition which might make the difference between losing or winning a project. As one respondent pointed out, 'so there is a stand-off. Everyone wants to collaborate in the ideal world but no one can because it involves giving away your competitive advantage'.

Another common problem is that collaboration with subcontractors often has to occur through union intermediaries and there are plenty of tensions here which can obstruct the innovation process. Views about the role of unions in the innovation process were typically polarized. Some firms have a confrontational approach to unions and see them as a barrier to innovation while others believe that it is easier to work with unions. Interactions between unions and construction firms are seen as highly politicized and in Australia changes in government bring large swings in political ideologies which produce a lack of continuity in the relationship which obstructs the innovation process. Despite the differing views about the role of unions in the innovation and their impact on productivity and competitiveness, most acknowledge that unions can play an important role through their close scrutiny of the industry: 'Unions ask the industry the difficult questions it won't ask itself and drive social innovations in areas such as safety which competition alone will not drive forward'.

Finally, as one contributor commented, 'the construction industry is a collection of other industries with different business models and production systems which generally don't fit'. For example, it is very hard for contractors to collaborate with steel companies because they cannot generally provide the long mills runs which are needed to make new products viable: 'This is where the term run-of-the-mill originated'. For the above reasons, meaningful supply chain collaborations around innovation are not easy to find. Collaborative innovation between competitors is even less common. However, one of the firms interviewed had collaborated with a leading architect to move his business from a contracting business model to a pre-fabricated building component manufacturing business. As he noted,

> We wanted to turn the industry on its head and by taking BIM to way ahead of anyone else in the construction industry and we have integrated our supply chain and processes so that everyone involved works off one model and creates a product in pre-kitted form . . . Working with manufacturing, which is far more mature and advanced in innovation because of their competitive pressures, we are transferring lessons from manufacturing into the rest of our business.

In another example of supply chain collaboration a multinational developer had developed a close relationship with a formwork subcontractor over many years to help fund, develop and implement a new formwork system which completes structures faster, safer and more cost competitively than other businesses. The same developer had collaborated with a prefabricated concrete manufacturer to incorporate mesh into concrete slab penetrations to drastically reduce the risk of falls on their sites. This is now used all around the world to reduce the risk of falls which is the greatest killer in the construction industry.

Organizing for innovation

There was general agreement that hierarchical bureaucratic organizations were more difficult for innovators than flat organic organizations:

> you have to go through more layers to get permission and the higher you go the more divorced from where the idea can add value. Top people are often isolated from reality and customers . . . top people also lose the sense they need to pick-up on an opportunity . . . This is why people leave and start up their own business. But successful big companies have processes to stop this.

While the idea of organic structures was recognizable to most of those interviewed, there was also widespread agreement that there was a danger of being too utopian here, in creating a world where business are too frightened to innovate. As one respondent argued, 'at the end of the day delivering projects is a highly pragmatic endeavour which requires detailed planning and strong accountability, discipline and reporting lines'. However, while there is a high degree of planning and control needed at a project level, there was also a strong sense of the need for decentralization at a business level. Innovative organizations seek to ensure that people are 'highly visible' and there are opportunities for anyone to contribute ideas to make the business better 'no matter what role they play in the business'. These companies seek to create an environment where there is a culture of talking and where people are empowered to solve problems by thinking outside the square and challenging the norms. In these firms, senior people are highly accessible and there is a chance for people with good ideas to 'figure out a role for themselves in the business' and to build new business opportunities around their personal interests. Another way that firms build an organic structure is to rotate people through different parts of the business. The project-based nature of the industry helps them to do this more easily. For example, one firm is structured into two simple groups: business development and delivery. The job of the business development group is to create new business opportunities whereas the job of the delivery group is to develop them. As projects start and finish, people continuously rotate in and out of these groups leading to a mixing and refreshing of experiences and insights across different projects.

Technology

Technology is becoming more important in the construction industry although many acknowledged that it is probably wise to be a 'fast follower' than a 'first mover' because those that first adopt new technologies often have to 'operate in a vacuum and pay for the costs of learning'. BIM was seen as a good example of this phenomenon. While there are a few companies that are experimenting with BIM, most of the industry is waiting for it to be proven. There are numerous stories of companies wasting a lot of time and resources in experimenting with BIM and being forced back into traditional ways of working to get projects finished on time: 'we are on our seventh BIM project and it's now cost neutral. There has been a lot of learning and risk. At some point in the future it will start to save us money.'

While in theory technologies like BIM can potentially facilitate greater innovation, there are many challenges in introducing such technologies into the construction sector. In particular, there are the long-established ways of working which have become institutionalized into the industry to change and then there are major problems in up-skilling the supply chain in using this technology. As one respondent noted,

> Construction is still fundamentally an industry based on relationships and people working together closely in a highly problem solving environment. So it is important to see technology not as a standalone piece but as part of a much larger organizational ecosystem and it generally won't work if it is treated in isolation from the rest of the business.

Catch it

Some felt that there is not always a catch-it phase. This is especially the case in industries like construction where innovations most often happen in response to a problem:

> a lot of innovation doesn't happen like this. Sometimes it's just solving a problem and you have a client. The catch-it analogy sounds too linear. It is really for the inventor who is running around with an idea looking for a buyer.

While this may be true, and while many innovations in construction arise in response to problems, there is also the view that the construction sector focuses too much on incremental operational improvements in neglect of the large strategic game-changers that can change the industry or provide them with competitive advantage. There was also the view that clients are generally reluctant to invest in innovations like this. So being able to sell an idea is important if the industry was to make a step-change: 'companies must better learn which client buttons to press'.

Most respondents also felt that this is the most critical and yet most difficult phase of the whole process, where many firms seem to struggle. It is seen as especially difficult in a small market like Australia compared to much larger regions like America and Europe which have their own internal markets. This means that many innovations 'fail at the final post'.

Some thought that very few firms have a strategic view of marketing which enables market intelligence and customer feedback to be fed back into business and innovation strategies. Furthermore, most companies don't have systems to systematically spot, assess and develop good ideas. So it often comes down the determination and passion of individuals to push their ideas through to reality. This ensures that many good ideas and people are lost from the industry: 'our sustainability strategy started with one person pushing it and never giving up. He was often knocked back but had a passion for it and eventually convinced everyone it was important'.

Where companies did have systematic processes they took many different forms which tended to reflect the business culture. For example, a number of companies had formalized the process through their intranet which allowed the whole business to vote on ideas and refine them through open dialogue. Those which are highly scored are then taken to an executive group where the best ones are chosen for further development and resourcing:

> In our business market research is crucial. We have a very strong reality-check. We make sure the numbers work at all stages of the process and we test the ideas quantitatively – dollars – and qualitatively – does it meet the needs and aspirations of our customers. We have a brilliant research team. Nothing is done without sign-off from research.

Uniquely, one company had a very informal approach where good ideas simple 'float to the surface' over time through a socialization process. This company described an active and vibrant web of communications throughout the business which picks up the good ideas. The culture was likened to the Aboriginal culture where few things were written down. Instead, ideas were transferred between generations in the form of stories: 'Ideas can circulate in this web for many years, until they get a sponsor who is near the centre of the web who can see an alignment with business goals and client needs. They then circulate at a higher level and become part of the conversations, acquire resources and become part of our language'.

Timing was also seen as critical to this phase of the innovation process. As one respondent said, 'unless there is a need at project level to either to win a bid or resolve a problem then it won't happen'. However, most acknowledged that there were very few 'willing' clients who were prepared to test and prototype a new idea on their project. And there are no methodologies for valuing innovation. Submitting a con-conforming bid was therefore a major risk. As one respondent said:

> There is always resistance, because new ideas involve change and the undoing of old systems and ways of doing things. Few people want to take the risk of trying something new and failing. There is no shortage of ideas in the industry but it is the opportunity for application which is often missing. It takes a courageous person to bring a new idea to fruition.

A good example of the importance of timing was a unique social innovation developed through an alliance with The Salvation Army. This involved using projects to help the long-term unemployed back to work through formal agreements with subcontractors. This innovation happened because of the confluence of three factors at a particular point in time: there was someone with the passion to drive it; the business leaders believed in it and supported it; and the Australian Government launched its Building Education Revolution programme to stimulate the industry in the midst of the global financial crisis and this programme had a social responsibility requirement.

Conclusion

The model of innovation presented in this chapter was the end result of reviewing the best international empirical research in the fields of innovation, business strategy and risk. By testing this model against the insights and experience of some of the industry's leading thinkers, it is clearer what can be achieved in reality. It is clear that there is no innovation formula and that anyone who purports to have found it has little understanding of what the process involves. Innovation is in essence a journey of self-reflection, driven by the desire to do better and its success is shaped and determined by a multitude of people and constantly changing factors which are often outside our control. The challenge of innovation is to recognize and harness this.

References

Abbot, C., Barrett, P., Ruddock, L. and Sexton, M. (2007) Hidden innovation in the construction and property sectors, *RICS Research Paper Series*, 7(20), December.

ABS (2006) *Census of Population and Housing, Cat no 2068.0*, Australian Bureau of Statistics, Canberra, Australia.

AFCC (1998) *Strategies for the Reduction of Claims and Disputes in the Construction Industry – A Research Report,* Australian Federation of Construction Contractors, Sydney, Australia.

Aghion, P. and Howitt, P. (1998) *Endogenous Growth Theory*, MIT Press, Cambridge, Massachusetts, USA.

Akintoye, A., Beck, M. and Hardcastle, C. (2003) *Public-Private Partnerships: Managing Risks and Opportunities*, Blackwell, Oxford, UK.

Akintoye, A. and Beck, M. (2009) *Policy, Finance and Management for Public-Private Partnership*, Blackwell, Oxford, UK.

Ansoff, H. I. (1965) *Corporate Strategy*, McGraw-Hill, New York.

Anthony, S. D. (2012) The new corporate garage, *Harvard Business Review*, September, 45–53.

APCC (1997) *Construct Australia: Building a Better Construction Industry in Australia*, Australian Procurement and Construction Council, Perth, APCC.

Argyris, C. (1999) *On Organizational Learning*, 2nd edn, Blackwell Publishing, Oxford, UK.

Ariely, D. (2009) *Predictably Irrational*, HarperCollins Publishers, Hammersmith, London.

Aronson, E. (1999) *The Social Animal*, 8th edn, Worth Publishers, New York.

Arora, A., Fosfuri, A. and Gambardella, A. (2001) *Markets for Technologies: The Economics of Innovation and Corporate Strategy*, MIT Press, Cambridge, Massachusetts, USA.

AS (2009) *Australian Labour Market Statistics January 2009*, catalogue no: 6105.5, Australian Bureau of Statistics, Canberra, Australia.

Asch, S. E. (1951) Effects of group pressure upon the modification and distortion of judgements, in H. Guetzkow (ed) *Groups, Leadership and Men*, Carnegie Press, Pittsburgh, PA, USA.

Atkinson, J. (1984) Manpower strategies for the flexible organization, *Personnel Management*, 6(1), 28–31.

Bandura, A. (1977) *Social Learning Theory*, General Learning Press, London.

Banwell, H. Sir (1964) *The Placing and Management of Contracts for Building and Civil Engineering*, HMSO, London.

Baregheh, A., Rowley, J. and Sambrook, S. (2009) Towards a multidisciplinary definition of innovation, *Management Decision*, 47(8), 1323–1339.

Baron, R. A. and Greenberg, J. (1992) *Behavior in Organizations: Understanding and Managing the Human Side of Work*, 3rd edn, Allyn and Bacon, Boston, USA.

Barrett, P. (2008) A global agenda for revaluing construction: the client's role, in P. Brandon and L. Shu-Ling (eds) *Clients Driving Innovation,* Wiley-Blackwell, Oxford, UK.

Barrett, P. S. and Lee, A. (2004) *Revaluing Construction: International Survey Questionnaire Results*, CIB, The Netherlands.

Barrett, P., Sexton, M. and Lee, A. (2008) *Innovation in Small Construction Firms*, Spon Press, London.

Basadur, M., Graen, G. B. and Scadura, T. A. (1986) Training effects on attitudes toward divergent thinking among manufacturing engineers, *Journal of Applied Psychology*, 71, 612–617.

Baumol, W. J. (2010) *The Microtheory of Innovative Entrepreneurship*, Princeton University Press, Princeton, USA.

BEC (1985) *A Fresh Look at the UK and US Building Industries,* The Building Employers Confederation, London.

Beck, U. (1992) *The Risk Society: Towards a New Modernity*, Sage Publications, Newbury Park, CA, USA.

Belbin, R. M. (1997) Conventional wisdom, *People Management*, 3(1), 36–38.

Bennett, P. (1999) Understanding responses to risk: some basic findings, in P. Bennett and K. Calman (eds) *Risk Communication and Public Health*, Oxford University Press, Oxford, UK, pp. 3–19.

Bennis, W. and O'Toole, J. (2005) How business schools lost their way, *Harvard Business Review*, 1 May, 96–99.

Berry, D. (2004) *Risk, Communication and Health Psychology*, Open University Press, Maidenhead, UK.

Best, R. (2012) International comparisons of cost and productivity in construction: a bad example, *Australasian Journal of Construction Economics and Building,* 12(3), 82–88.

Betts, M. and Ofori, G. (1992) Strategic planning for competitive advantage, *Construction Management and Economics*, 10(3), 511–532.

BPF (1984) *Manual of the BPF system – The British Property Federation System for Building Design and Construction,* The British Property Federation Ltd., London, UK.

Braben, D. W. (2004) *Pioneering Research – A Risk Worth Taking,* John Wiley and Sons, Hoboken, NY, USA.

Brandon, P. (2010) Rise of machines, *RICS Business Journal,* January, 29–30.

Brandon, P. and Shu-Ling, L. (2008) *Clients Driving Innovation*, Wiley-Blackwell, Oxford, UK.

Brass, D. J. (2003) A social network perspective on human resources management, in R. Cross, A. Parker and L. Sasson (eds) *Networks in the Knowledge Economy*, Oxford University Press, New York.

Bresnen, M. (2008) Reifying the client in construction management research, in P. Brandon and L. Shu-Ling (2008) *Clients Driving Innovation*, Wiley-Blackwell, Oxford, pp. 47–57.

Bresnen, M. and Marshall, N. (2000) Partnering in construction: a critical review of issues, problems and dilemmas, *Construction Management and Economics*, 18, 229–237.

Brown, M. (2005) *Corporate Integrity: Rethinking Organizational Ethics and Leadership*, Cambridge University Press, Cambridge.

Bryan, L.L., Matson, E. and Weiss, L. M. (2007) Harnessing the power of informal employee networks, *The McKinsey Quarterly Organization*, 4, 13–19.

Bungay, S. (2010) What the Prussians could do for us, *The Sunday Times*, 5 December, 4–5.

Burnes, B. (2004) Kurt Lewin and the planned approach to change: a re-appraisal, *Journal of Management Studies*, 41(6), 977–1002.

Burns, T. and Stalker, G. M. (1961) *The Management of Innovation*, Tavistock, London.

Burt, R. S. (2000) The network structure of social capital, *Research in Organizational Behavior*, 22(2), 134–157.

Burt, R. S. (2005) *Brokerage and Closure: An Introduction to Social Capital*, Oxford University Press, New York.

Cai, P. (3013) Business can benefit from students of humanities, *The Sydney Morning Herald*, 28 January, 1–2.

Cambourne, K. (2007) Turning a dollar from research, *The Sydney Morning Herald*, 1 October, 15–16.

Camillus, J. C. (2008) Strategy as a wicked problem, *Harvard Business Review*, 86(5), 98–106.

Camp, R. C. (1989) *Benchmarking: The Search for Industry Best Practices that Lead to Superior Performance*, ASQC Quality Press, Milwaukee, WI.

Carthey, J. (2007) Healthcare designers and information use, *ConnectED International Conference on Design Education 2007*, Sydney, Australia, 9–12 July, 88–98.

Castells, M. (1996) *The Rise of the Networked Society: The Information Age*, Blackwell Publishers, Cambridge; Massachusetts, USA.

Chan, P. and Cooper, R. (2011) *Constructing Futures: Industry Leaders and Futures Thinking in Construction*, Wiley Blackwell, Chichester, UK.

Chandler, A. (1962) *Strategy and Structure*, MIT Press, Cambridge, MA, USA.

CII (2008) *Leveraging Technology to Improve Construction Productivity*, Construction Industry Institute, New York.

Clegg, S., Carter, C., Kornberger, M. and Schweitzer, J. (2011) *Strategy – Theory and Practice*, Sage, London.

COA (2009) *Powering Ideas: An Innovation Agenda for the 21st Century*, Commonwealth of Australia, Canberra, Australia.

Cohen, J. D. and Schooler, J. W. (eds) (1996) *Scientific Approaches to Consciousness*, Carnegie Mellon Symposia on Cognition Series, Psychology Press, New York, USA.

Collins, J. C. and Porras, J. I. (1994) *Built to Last; Successful Habits of Visionary Companies*, Random House, Australia.

Cormier, D. (2011) *Rhizomatic Education: Community as Curriculum*, available online http://www.innovateonline.info/index.php?view=article&id=550, (accessed 10 July 2011).

Csikszentmihalyi, M. (1996). *Creativity: Flow and the Psychology of Discovery and Invention*, Harper Perennial, New York, USA.

CSIRO (2010) *Our Future World*, CSIRO Development Group, Melbourne, March 2010.

CSS (1990) *Investing in Building*, Centre for Strategic Studies, University of Reading, UK.

Cummings, C. and Daellenbach, U. (2009) A guide to the future of strategy? The history of long range planning, *Long Range Planning*, 42, 234–263.

Cummings, S. (2002) *Recreating Strategy*, Sage, London.

Cutler, T. (2008) *Venturousaustralia*, Department of Innovation, Industry, Science and Research, Commonwealth Government, Canberra.

Dainty, A. and Loosemore, M. (2012) *Human Resource Management in Construction: Critical Perspectives*, Routledge, London.

De hann, J., Voordijk, H. and Joosten, G. J. (2002) Market strategies and core capabilities in the building industry, *Construction Management and Economics*, 20(10), 109–118.

de Man, A., Berends, H., Lammers, I., van Raaij, E. and Van Weele, A. (2008) Knowledge and innovation in networks, a conceptual framework, in A. de Man (ed.) *Knowledge Management and Innovation in Networks*, Edward Elgar, Cheltenham, UK, pp. 1–14.

De Valence, G. (1997) *Construction Industry Reform Strategies in Australia: CIDA and the CPSC*, First International Conference on Construction Industry Development, National University of Singapore, 124–131.

Deal, T. E. and Kennedy, A. A. (1982) *Corporate Cultures: The Rites and Rituals of Corporate Life*, Addison-Wesley, New York.

DIISR (2010) *Australian Innovation System Report 2010*, Department of Innovation, Industry, Science and Research, Commonwealth Government of Australia, Canberra, Industry Innovation Councils Team Manufacturing Division.

Dimaggio, P. J. and Powell, W. (1983) The iron cage revisited: institutional isomorphism and collective rationality in organizational fields, *American Sociological Review*, 48, 147–160.

DIR (1989) *Review of the Structural Efficiency Principle*, February–May, Department of Industrial Relations, Commonwealth Government, Australian Government Publishing Service, Canberra.

DISR (1999) *Building for Growth*, Department of Industry, Science and Resources, Canberra.

DITAC (1989), *Indicative Survey of Non-Residential Construction Industry Efficiency*, Department of Industry, Technology and Commerce, Canberra.

Dixon, T., Pottinger, G. and Jordan, A. (2005) Lessons from the private finance initiative in the UK: benefits, problems and critical success factors, *Journal of Property Investment & Finance*, 23(5), 412–423.

Dodgson, M. (2008) Commerce in fresh ideas, *The Australian*, 17 September, 30–31.

Dodgson, M. and Gann, D. (2010) *Innovation: A Very Short Introduction*, Oxford University Press, Oxford, UK.

Dodgson, M. and Hinze, S. (2000) Indicators used to measure innovation process: defects and possible remedies, *Research Evaluation*, 8(2), 101–114.

Dodgson, M., Gann, D. and Salter, A. (2005), *Think, Play, Do – Technology, Innovation, and Organization*, Oxford University Press, Oxford.

DPWS (1998) *Construct NSW*, NSW Department of Public Works and Services, Sydney, NSW.

Driskill, G. W. and Brenton, A. L. (2011) *Organisational Culture in Action*, 2nd edn, Sage, London.

Duffy, F. and Powell, K. (1998) *The New Office*, Conrad Octopus, New York.

Dumbelton, J. H. (1986) *Management of High Technology Research and Development*, Elsevier Science Publishing Company, The Netherlands.

Dunn, K. M. and McDonald, A. (2001) The geography of racisms in NSW: a theoretical exploration and some preliminary findings, *Australian Geographer*, 32(1), 29–44.

Edkins, A. and Winch, G. (1999) *The Performance of the UK Construction Industry: An International Perspective*, Engineering and Physical Sciences Research Council's Construction as a Manufacturing Process' research programme, EPSRC, London, UK.

Egan, J. Sir (1998) *Rethinking Construction*, Department of Trade and Industry, HMSO, London.

Egan, K. (1992) *Imagination in Teaching and Learning: Ages 8–15,* Routledge, London.

Egbu, C. (2008) Clients' roles and contributions to innovations in the construction industry: when giants learn to dance, in P. Brandon and L. Shu-Ling (eds) *Clients Driving Innovation*, Wiley-Blackwell, Oxford, pp. 69–77.

Eisenhardt, K. M. and Martin, J. A. (2000) Dynamic capabilities: what are they? *Strategic Management Journal,* 21(10–11), 1105–1121.

Eisenstat, R., Foote, N., Galbraith, J. and Miller, D. (2001) Beyond the business unit, *The McKinsey Quarterly*, 1, January, 5–9.

Ekvall, G. (1999) Creative climate, in M. A. Runco and S. R. Pritzker (eds) *Encyclopedia of Creativity, vol. 1 A–H,* Academic Press, San Diego, pp. 403–412.

Emmerson, H. Sir (1962) *Survey of Problems before the Construction Industries*, HMSO, London.

EOWWA (2012) *Women in the Workforce by Industry, Equal Opportunity for Women in the Workplace Agency (Australia)*, Australian Government, Canberra.

Ernst and Young (2009) *Entrepreneurship and Innovation*, Ernst and Young, London.

Eslake, S. and Walsh, M. (2011) *Australia's Productivity Challenge*, Grattan Institute, Melbourne, Australia.

EUCO (1995) *The Leven Scrutiny*, Efficiency Unit, Cabinet Office, HMSO, London.

Fairclough, J. Sir (2002) *Rethinking Construction Innovation and Research; A Review of Government Policies and Practices*, Department of Trade and Industry, London, UK.

FCAPP (2001) *Constructing the Future*, Foresight-Construction Associate Programme Panel, Department of Trade and Industry, London.

Ferguson, A. (2012) How the humble app is changing, *The Sydney Morning Herald*, Weekend Business, 19–20 May, 6–7.

Festinger, L. (1957) *A Theory of Cognitive Dissonance*, Stanford University Press, Stanford, USA.

Flyvbjerg, B. (2005) *Policy and Planning for Large Infrastructure Projects: Problems, Causes, Cures*, World Bank Policy research working paper 3781, World Bank, Infrastructure Network, Office of the Vice President, December, CA, USA.

Flyvberg, B., Bruzelius, N. and Rothengatter, W. (2003) *Megaprojects and Risk: An Anatomy of Ambition,* Cambridge University Press, Cambridge, UK.

Foreman, P. D. and Whetten, A. (2002) Members' identification with multiple-identity organizations, *Organizational Science,* 13(6), 618–635.

Fox, C. and Hooper, N. (2011) Test of mettle, *AFRBOSS Magazine*, 20–28.

Freeman, L. C. (1979) Centrality in social networks conceptual clarification, *Social Networks*, 1(2), 215–239.

Freeman, L. C. (2010) What is Social Network Analysis?, available online http://www.insna.org/sna/what.html (accessed 2 April 2010).

French, J. R. P. and Raven, B. (1959) The bases of social power, in D. Cartwright and A. Zander, *Group Dynamics*, Harper & Row, New York, pp. 78–94.

Freud, S. (1921) Group psychology and the analysis of the ego, in J. Strachey (ed.) *Standard Edition of the Complete Psychological Works*, vol 18, Hogarth Press, London.

Freud, S. (1954) *The Interpretation of Dreams*, Allen and Unwin, London.

Friedman, J. L. (2005) *The World Is Flat*, Penguin Books, London.

Fuglsang, L. (2008) Innovation with care: what it means, in L. Fuglsang (ed.) *Innovation and the Creative Process*, Edward Elgar, Cheltenham, UK, chapter 1, pp. 3–25.

Furnhman, A. and Gunter, B. (1994) *Business Watching*, ABRA Press, London.

Gambatese, J. A. and Hallowell, M. (2011) Factors that influence the development and diffusion of technical innovations in the construction industry, *Construction Management and Economics*, 29(4), 507–517.

Gann, D. M. (2000) *Building Innovation – Complex Constructs in a Changing World*, Thomas Telford, London.

Gann, D. M. (2003) Guest editorial: innovation in the built environment, *Construction Management and Economics*, 21(4), 553–555.

Gardner, H. (1993) *Creating Minds*, Basic Books, New York.

Garling, P. (1991) *Research and Development – Preparing for the Next Building Boom: The Project Deliverer's View*, Building Science Forum of Australia, Sydney, Australia.

Gigerenzer, G. (2003) *Reckoning with Risk*, Penguin, London.

Granovetter, M. (1973) The strength of weak ties, *The American Journal of Sociology*, 78(6), 1360–1380.

Grant, R. M. (1996) Towards a knowledge-based theory of the firm, *Strategic Management Journal,* 17(Winter Special Issue), 109–122.

Graves, D. (1986) *Corporate Culture, Diagnosis and Change*, Frances Pinter, London.

Green, R., Toner, P. and Agarwal, R. (2012) *Understanding Productivity – Australia's Choice*, University of Technology, Sydney.

Green, S. D. (1998) The technocratic totalitarianism of construction process improvement: a critical perspective, *Engineering, Construction and Architectural Management*, 5(4), 376–386.

Green, S. D. (1999) The missing arguments of lean construction, *Construction Management and Economics*, 17, 133–137.

Green, S. D. (2006) Discourse and fashion in supply chain management, in S. Pryke and H. Smyth (eds) *The Management of Complex Projects*, Blackwell Publishing, Oxford, UK, pp. 236–251.

Green, S. D. (2009) The evolution of corporate social responsibility in construction, in M. Murray and A. Dainty (eds) *Corporate Social Responsibility in the Construction Industry*, Taylor and Francis, London, pp. 24–54.

Green, S. D. (2011) *Making Sense of Construction Improvement*, Wiley Blackwell, Oxford, UK.

Green, S. D., Larsen, G. D. and Kao, C. C. (2008) Competitive strategy revisited: contested concepts and dynamic capabilities, *Construction Management and Economics*, 26, 63–78.

Greenstein, G. (1998) *Portraits of Discovery*, John Wiley and Sons Inc, New York, USA.

Gronroos, C. (2000) *Service Management and Marketing*, 2nd edn, Wiley, Chichester, UK.

Haber, R. N. (1970) How we remember what we see, *Scientific American*, 222, 104–112.

Hagel, J., Seeley Brown, J. and Davison, L. (2010) *The Power of Pull: How Small Moves, Smartly Made, Can Set Big Things in Motion*, Badic Books, New York, USA.

Hall, B. H. (2011) *Using Productivity Growth as an Innovation Indicator, Report for the High Level Panel on Measuring Innovation*, D G Research, European Commission, University of Maastricht and UC Berkeley, USA.

Hamel, G. (2002) *Leading the Revolution*, Plume, New York.

Handy, C. (1985) *Understanding Organizations,* Harmondsworth, New York, Penguin.

Hannaford, K. (2010) Our miracle material, *The Daily Telegraph*, 18 September, 12–14.

Harty, C., Goodier, C. I., Soetanto, R., Austin, S., Dainty, A. and Price, A. (2007) The futures of construction: a critical review of construction future studies, *Construction Management and Economics*, 25(3), 477–493.

Hawksworth, J. and Cookson, G. (2008) *The World in 2050: Beyond BRICS, a Broader Look at Emerging Market Growth Prospects*, PricewaterhouseCoopers, Sydney, Australia.

He, Z. and Wong, P. (2004) Exploration vs. exploitation: an empirical test of the ambidexterity hypothesis, *Organisation Science,* 15(4), 481–494.

Heller, R. (1993) TQM – not a panacea but a pilgrimage, *Management Today,* January, 37–40.

Higgin, J. and Jessop, N. (1965) *Communications in the Building Industry*, Tavistock, London.

Hillebrandt, P. M. (1984) *Analysis of the British Construction Industry*, Macmillan, London

Hillebrandt, P M. and Cannon, J. (eds) (1989) *The Management of Construction Firms: Aspects of Theory*, Macmillan, Basingstoke, UK.

Hillebrandt, P. and Cannon, J. (1994) *The Modern Construction Firm*, Macmillan, London.

Hilmer, F. and Field, L. (2011) Effective research must be for the long haul, *Australian Financial Review*, 4 May, 63–64.

Hilmer, F. G. and Donaldson, L. (1996) *Management Redeemed – Debunking the Fads That Undermine Corporate Performance,* The Free Press, London.

Hobson, J. and Treadway, K. (2008) Is the client really part of the team? A contemporary policy perspective on Latham/Egan, in P. Brandon and L. Shu-Ling (eds) *Clients Driving Innovation,* Wiley-Blackwell, Oxford, UK.

Hofstede, G. (1980), *Cultural Consequences: International Differences in Work Related Values*, Sage Publications, Beverly Hills, California, USA.

Hooper, B. and Harris, M. (2010) 2020 Vision, *Modus*, Royal Institution of Chartered Surveyors, London, UK, 34–36.

Hornstein, H. A. (1987) *Managerial Courage*, John Wiley and Sons, New York.

Horrigan, D. (2011) *Strategic Serendipity: The Art of Being in the Right Place at the Right Time . . . with the Right People*, September 2011, Australia Business Foundation, Sydney, Australia.

Hubbard, G., Samuel, D., Heap, S. and Cocks, G. (2002) *The First XI: The Winning Organisations in Australia*, Wiley, London.

Huczynski, A. A. (1993) Explaining the succession of management fads, *The International Journal of Human Resource Management*, 4(2), 443–463.

ICE (1998) *RAMP*, The Institute of Civil Engineers and the Faculty and Institute of Actuaries, Thomas Telford, London.

IGT (2010) *Low Carbon Construction, Innovation and Growth Team*, HMSO, London.

Imber, A (2011) *The Creativity Formula: 50 Scientifically Proven Creativity Boosters For Work And For Life*, Liminal Press, Australia.

Ivory, C. (2005) The cult of customer responsiveness: is design innovation the price of a client-focused construction industry? *Construction Management and Economics*, 23(8), 861–870.

Jacobson, M. Z. and Delucchi, M. A. (2009) A path to sustainable energy by 2030, *Scientific American*, November, 58–65.

Janis, I. L. (1988) Groupthink, in J. B. Lau and A. B. Shani (eds) *Behavior in Organizations – An Experiential Approach*, 4th edn, Irwin Homewood, Urbana, Illinois, pp. 162–169.

Johnson, J. and Tatum, C. B. (1993) Technology in marine construction firms, *ASCE Journal of Construction Engineering and Management,* 119(1), 148–162.

Kahneman, D. (2011) *Thinking, Fast And Slow*, Penguin, New York.

Kahneman, D., Tversky, A. and Slovic, P. (eds) (1982) *Judgment under Uncertainty: Heuristics & Biases*, Cambridge University Press, Cambridge, UK.

Kanter, E. (1983) *The Change Masters*, Allen and Unwin, London.

Kaplan, A. (2004) *The Conduct of Inquiry*, 4th edn, Transaction Publishers, New Jersey, New Brunswick, USA.

Kaplan, R. S. and Norton, D. P. (1992) The balanced scorecard – measures that drive performance, *Harvard Business Review*, 69(1), 71–79.

Kasperson, R., Renn O., Slovic, P., Brown, H., Emel, J., Goble, R. and Ratick, S. (1988) The social amplification of risk: a conceptual framework, *Risk Analysis*, 8(1), 177–187.

Katz, S. N. (2002) The path breaking, fractionalized, uncertain world of knowledge, *The Chronicle of Higher Education*, 20 September.

Kay, J. (1993) *Foundations of Corporate Success: How Business Strategies Add Value*, Oxford University Press, Oxford, UK.

Kay, J. (2013) The lesson of Victorian London's rise from sewer to spectacle, *Financial Times*, 16 January, 9–10.

Kerin, J. and Cook, P. (1989) *Research, Innovation and Competitiveness*, AGPS, Canberra.

King, N. (1992) Modeling the innovation process: an empirical comparison of approaches, *Journal of Occupational and Organizational Psychology*, 65(1), 89–100.

Kipping, M. and Engwall, L. (eds) (2002) *Management Consulting Emergence and Dynamics of a Knowledge Industry*, Oxford University Press, Oxford, UK.

Kotler, P. and Lee, N. (2005) *Corporate Social Responsibility*, Wiley, Hoboken, New Jersey

Koubatis, A. and Schonberger, J. Y. (2005) Risk management in complex systems, *International Journal of Critical Infrastructures*, 1(2/3), 195–215.

Lafley, A. G., Martin, R. L., Rivkin, J. W. and Siggelkow, N. (2012) Bringing science to the art of strategy, *Harvard Business Review*, 89(1), 56–66.

Lamborde, M. and Sanvido, V. (1994) Introducing new technologies into construction companies, *ASCE Journal of Construction Engineering and Management*, 120(3), 334–345.

Langford, D. and Male, S. (2001) *Strategic Management in Construction*, Blackwell, Oxford, UK.

Lansley, P. (1981) Corporate dislocation: a threat for the 1980s, *Journal of General Management,* 9(3), 399–416.

Lansley, P., Quince, T. and Lea, E. (1979) *Flexibility and Efficiency in Construction Management*, Building Industry Group, Ashridge Management College, Amersham, Bucks, UK.

Latham, M. Sir (1994) *Constructing the Team*, HMSO, London.

Le Bon, G. (1908) *The Crowd: A Study of the Popular Mind*, Unwin, London.

Le Masson, P., Weil, B. and Hatchuel, A. (2010) *Strategic Management of Innovation and Design*, Cambridge University Press, Cambridge, UK.

Leiringer, R. (2006) Technological innovations in PPPs: incentives, opportunities and actions, *Construction Management and Economics*, 24(3), 301–308.

Lenard, D. J. (1996*) Innovation and Industrial Culture in the Australian Construction Industry: A Comparative Benchmarking Analysis of Critical Cultural Indices Underpinning Innovation*, unpublished PhD Thesis, University of Newcastle, Newcastle, Australia.

Lewin, K. (1947) Frontiers in group dynamics, *Human Relations*, 6(1), 2–38.

Lieberman, M. and Montgomery, D. B. (1988a) First-mover advantages, *Strategic Management Journal*, 9, 41–58.

Lieberman, M. B. and Montgomery, D. B. (1998b) First-mover (dis)advantages: retrospective and link with the resource-based view, *Strategic Management Journal* 19(12), 1111–1125.

Liker, J. K., Haddad, C. V. and Karlin, J. (1999) Perspectives on technology and work organization, *Annual Review of Sociology*, 25(4), 575–596.

Lim, B and Ofori, G (2007) Classification of innovation for strategic decision making in construction businesses, *Construction Management and Economics*, 25(9) 963–979.

Locke, R. R. and Spender, J.-C. (2011) *Confronting Managerialism*, Zed Books, London and New York.

Lofthouse, J. A. (1965) *Efficiency in Road Construction*, HMSO, London.

Loosemore, M. (1999) Power, responsibility and construction conflict, *Construction Management and Economics*, 17(6), 699–711.

Loosemore, M. and Holliday, S. (2012) Pro-innovation bias in construction, *Association of Researchers in Construction Management (ARCOM): 28th Annual Conference,* Edinburgh, United Kingdom, 3–5 September 2012, pp. 787–797.

Loosemore, M. and Phua, F. (2011) *Corporate Social Responsibility in the Construction Industry: Doing the Right Thing?* Routledge, London, UK.

Loosemore, M., Choo, H. and Koh, J. (2002) Encouraging R&D in construction, *ASCE Journal of Professional Issues in Engineering Education and Practice*, 128(1), 25–30.

Loosemore, M., Chow, V. W and McGeorge, D. (2012) Modelling the risks of extreme weather events to Australasian hospital infrastructure using rich picture diagrams, *Construction Management and Economics,* 30(12), 1071–1086.

Loosemore, M., Dainty, A. and Lingard, H. (2003) *Human Resource Management in Construction Projects: Strategic and Operational Approaches*, Taylor and Francis, London, UK.

Loosemore, M., Phua, F., Dunn, K. and Ozguc, U. (2010), Operatives' experiences of cultural diversity on Australian construction sites, *Construction Management and Economics*, 28(2), 173–185.

Loosemore, M., Raftery, J., Reilly, C. and Higgon, D. (2005) *Risk Management in Projects*, 2nd edn, Spon Press, London, UK.

Lundvall, B. A. (2010) *National Systems of Innovation: Toward a Theory of Innovation and Interactive Learning*, Arnold, London.

Mac Gibbon, A. (2010) Time out boosts brains, *Sydney Morning Herald*, 2 August, 15–16.

MacLeod, D. and Clarke, N. (2010) *Engaging for Success*, Office of Public Sector Information, UK Government, London, UK.

Maddock Report (1989) *The Predictability of Australian Project Performance*, Department of Industry, Technology and Resources, Victoria.

Male, S. P. and Stocks, R. K. (eds) (1991) *Competitive Advantage in Construction*, Butterworth Heinemann, Oxford, UK.

Manley, K. (2006) The innovation competence of repeat public sector clients in the Australian construction industry, *Construction Management and Economics*, 24(2), 1295–1304.

Manley, K. and Rose, T. (2009) *Innovative Practices in the Australian Built Environment Sector: An Information Resource for Industry*, Cooperative Centre for Construction Innovation, Queensland University of Technology, Australia.

Manning, P. (2011) Peak oil: it's closer than you think, *Sydney Morning Herald*, 30 April, 16–17.

March, J. G. (1999) *The Pursuit of Organizational Intelligence*, Blackwell, Oxford, UK.

Marosszeky, M. (1999) Technology and Innovation. in R. Best and G. de Valence (eds) *Building in Value*, Arnold, London.

Mason, M. M., Norton, M. I., Van Horn, J., Wegner, D. M., Grafton, S. T. and

Macrae, C. N. (2007) Wandering minds: the default network and stimulus-independent thought, *Science*, 19, 315(5810), 393–395.

Mason, R. (2012) Taxpayers fund £150bn bail-out for 21 PFI hospitals at severe risk, *The Daily Telegraph*, 4 February, 10–11.

Matthews, M. (2008) Some of the exciting ideas we need, *The Australian*, Higher Education, 11 June, 25–26.

McGeorge, D. and Loosemore, M. (2003) *An Investigation into Productivity and Reform in the International Building and Construction Industry*, Discussion Paper 6, Royal Commission into the Australian Building and Construction Industry, Melbourne, Victoria, Australia.

McKeown, M. (2012), *The Strategy Book*, FT Prentice Hall, London.

McKinsey (2005) Views from the boardroom, *McKinsey Quarterly*, January.

Mendonca, S., Cunha, M. and Clegg, S. (2008) Unsought innovation: serendipity in organizations, paper presented at the 25th Celebration Conference on *Entrepreneurship and Innovation – Organizations, Institutions, Systems and Regions*, Copenhagen, CBS, Denmark, 17–20 June, pp. 348–359.

Merton, R. K. and Barber, E. G. (2003) *The Travels and Adventures of Serendipity: A Study in Sociological Semantics and the Sociology of Science*, Princeton University Press, Princeton, USA.

Mills, K. (2012) BRW thirty most innovative companies, *BRW*, 6 December, 20–35.

Mims, C. (2011) *The City Of The Future Will Be Covered In Lichen,* Fast Company, 28 November, available online http://www.fastcompany.com/biomimicry/combating-urban-heat-by-growing-lichen-on-buildings (accessed 24 October 2012).

Mitchell, H. (2012) Fostering creative communities key to escaping limiting island mindset, *The Sydney Morning Herald*, 10–11 November, Weekend Business, 8–9.

Minztberg, H. (1984) *The Nature of Managerial Work*, 2nd edn, Prentice Hall, Englewood Cliffs, NJ, USA.

Morgan, M. G. and Keith, D. W. (1995) Subjective judgements by climate experts, *Environmental Science and Technology*, 29(3), 468–476.

Morris, L. (2008) *Innovation Metrics: The Innovation Process and How to Measure It*, an InnovationLabs White Paper, November 2008, InnovationLabs LLC, California, USA.

Muller, T. and Becker, L. (2012) *Get Lucky: How to Put Planned Serendipity to Work for You and Your Business*, John Wiley & Sons, New York, USA.

Muir, C. (2000) *Achieving Success Through Social Capital*, Jossey-Bass Inc., San Francisco.

Murray, M. and Langford, D. (2003) *Construction Reports: 1944–98*, Blackwell, Oxford, UK.

Nam, C. N. and Tatum, C. B. (1992) Strategies for technology push – lessons from construction innovations, *ASCE Journal of Construction Engineering and Management*, 18(3), 213–234.

Nam, C. N. and Tatum, C. B. (1997) Leaders and champions for construction innovation, *Construction Management and Economics*, 15(3), 259–270.

NAO (2001) *Modernising Construction*, National Audit Office, London.

NEDO (1983) *Faster Building for Industry*, National Economic Development Office, HMSO, Millbank, London.

NEDO (1988) *Faster Building for Commerce*, National Economic Development Office, HMSO, Mill Bank, London

Ness, K. (2010) The discourse of 'Respect for People' in UK construction, *Construction Management and Economics*, 28(5), 481–493.

Ness, K, and Green, S. (2012) Human resource management in the construction context: if only we could find those workers, in A. Dainty and M. Loosemore (eds) *HRM in Construction: Critical Perspectives*, Routledge, London.

NESTA (2007) *Hidden Innovation: How Innovation Happens in Six 'Low Innovation' Sectors*, National Endowment for Science, Technology and the Arts, London, UK.

NESTA (2009) *The Connected University: Driving Recovery and Growth in the UK Economy*, National Endowment for Science, Technology and the Arts, London, UK.

Newcombe, B. (1976) *The Evolution and Structure of the Construction Firm*, unpublished MSc thesis, University College London.

Nonaka, I. (1991) The knowledge creating company, *Harvard Business Review*, November–December, 96–104.

Nonaka, I. and Takeuchi, H. (1995) *The Knowledge-creating Company: How Japanese Companies Create the Dynamics of Innovation*, Oxford University Press, Oxford, UK.

OECD (2002) *Frascati Manual – Proposed Standard Practice for Surveys on Research and Experimental Development*, 6th edn, Organisation for Economic Cooperation and Development, Paris, France.

OECD (2005) *Oslo Manual – Guidelines for Collecting and Interpreting Innovation Data,* Organisation for Economic Cooperation and Development, Paris, France.

OECD (2010) *Measuring Innovation: A New Perspective*, Organisation for Economic Cooperation and Development, Paris, France.

Oliver, J. (1993) Shocking to the core, *Management Today*, August, 18–22.

Oldfield, S. (2012) Still a lucky country, despite rise and fall, affirms Blainey, *The Sydney Morning Herald*, 10 Wednesday October, 9–10.

Orr, G. and Roth, E. (2012) A CEO's guide to innovation in China, *McKinsey Quarterly*, February, 5–8.

Oury, J. (2007) *A Guide to Corporate Social Responsibility*, British Standards Institution, London.

Palangkaraya, A., Stierwald, A., Webster, E. and Jenson, P. H. (2010) *Examining the Characteristics of Innovative Firms in Australia*, Department of Innovation, Industry, Science and Research, Commonwealth Government, Canberra, Australia.

Panati, C. (1991) *Panati's Parade of Fads, Follies and Manias – The Origins of Our Most Cherished Obsessions*, Harper Perennial, Park, New York, USA.

Parnes, S. J., Noller, R. B. and Biondi, A. M. (1977) *Guide to Creative Action*, Scribner, New York.

Pascale, R. T. (1991) *Managing on the Edge*, Penguin Books, Harmondsworth, UK.

Paulus, P. B., Dzindolet, M. T., Poletes, G. and Camacho, L. M. (1993) Perception of performance in group brainstorming: the illusion of group productivity, *Personality and Social Science Bulletin*, 19, 78–89.

Pedersen, A., Walker, I. and Wise, M. (2005) Talk does not cook rice: beyond anti-racism rhetoric to strategies for social action, *Australian Psychologist*, 40(1), 20–30.

Peters, T. J. (1989) *Thriving on Chaos*, Macmillan, London.

Peters, T. J. and Waterman, R. H. Jnr. (1982) *In Search of Excellence*, Harper and Row, New York.

Pettigrew, A., Thomas, H. and Whittington, R. (eds) (2006) *Handbook of Strategy and Management*, Sage, London.

Porter, M. E. (1980) *Competitive Strategy: Techniques for Analyzing Industries and Competitors*, Free Press, New York.

Porter M. E. (1985) *Competitive Advantage: Creating and Maintaining Superior Performance*, Free Press, New York.

Porter, M. E. (1987) *Corporate Strategy: The State of Strategic Thinking,* Free Press, New York.

Porter, M.E. (1998) *The Competitive Advantage of Nations,* Free Press, New York.

Porter, M. E. and Kramer, M. R. (2006) Strategy and society: the link between competitive advantage and corporate social responsibility, *Harvard Business Review*, 86, 78–94.

Porter, M. E., Hills, G., Pfitzer, M., Patscheke, S. and Hawkins, E. (2011) *Measuring Shared Value*, FSG, New York.

Prahalad, C. K. and Bettis, R. A. (1986) The dominant logic: a new linkage between diversity and performance, *Strategic Management Journal*, 7(6), 485–501.

Prahalad, C. K. and Hamel, G. (1990) The core competence of the corporation, *Harvard Business Review*, 90, 79–91.

RCBCI (1992) *Final Report* by R. V. Gyles QC, Vols. 1–10, Royal Commission into Productivity in the Building Industry in NSW, The Commission, Sydney, Australia.

RCBCI (2002a) *Final Report*, Royal Commission into the Building and Construction Industry, Melbourne, Victoria, Australia.

RCBCI (2002b) *Overview of the Nature and Operation of the Building and Construction Industry*, Royal Commission into The Building and Construction Industry, Melbourne, Victoria, Australia.

RCBCI (2002c) *Productivity and Performance in the Building and Construction Industry*, Royal Commission into The Building and Construction Industry, Melbourne, Victoria, Australia.

Reed, H. and Loosemore, M. (2012) Culture shock of alliance projects, in *Association of Researchers in Construction Management (ARCOM): 28th Annual Conference,* Edinburgh, United Kingdom, 3–5 September, 543–553.

Richardson, B. (1996) Modern management's role in the demise of a sustainable society, *Journal of Contingencies and Crisis Management*, 4(1), 20–32.

Richmond, S. (2012) Smile for the smartphone, *The Daily Telegraph*, 21 January, 40–41.

Risk (2009) More regulation a worry for in-house counsel, *Riskmagazine,* April, 12–13.

Robbins, S. P., Bergman, R., Stagg, I. and Coulter, M. (2003) *Foundations of Management*, Prentice Hall, Sydney, Australia.

Rogers, E. M. (1962) *Diffusion of Innovations*, Glencoe: Free Press, New York.

Rogers, E. M. (1995) *Diffusion of Innovations*, 4th edn, Free Press, New York.

Rogers, E. M. and Kincaid, D. L. (1981) *Communication Networks. Toward a New Paradigm for Research.* Free Press, New York.

Rosen, C. (2010) Out of step with time, *Sydney Morning Herald*, 9–10 October, 6–7.

Rosenthal, J. S. (2005) *Struck by Lightning – The Curious World of Probabilities*, Harper Collins Publishers, Ontario, Canada.

Rotov, I. and Baron, J. (1990) Reluctance to vaccinate: omission bias and ambiguity, *Journal of Behavioural Decision Making*, 3(2), 263–77.

Rowe, G. and Wright, G. (2001) Differences in experts and lay judgments of risk: myth or reality, *Risk Analysis*, 21, 341–56.

Rushton, T. (2010) Opening Pandora's box, *Building Surveying Journal*, July–August, RICS, London, 26–30.

Ryan, K. D. and Oestreich, D. K. (1998) *Driving Fear out of the Workplace*, 2nd edn, Jossey-Bass Publishers, San Francisco.

Sachs, J. D. (2012) A world adrift: global economy facing sustainability crisis, *The Jakarta Post*, 27 April, 5–6.

Sage, D. J. and Dainty, A. R. J. (2012) Understanding power within project work: the neglected role of material and embodied registers, *Engineering Project Organization Journal*, 2(4), 202–215.

Salter, A. and Gann, D. (2005) Last among equals: a comparison of innovation in construction, services and manufacturing in the UK, *Construction Management and Economics*, 21(6), 603–612.

Samson, D. (2011) *Innovation for Business Success: Achieving a Systematic Innovation Capability*, Department of Innovation, Industry, Science and Research, Commonwealth Government, Canberra, Australia.

Santrock, J. W. (1998) *Psychology*, 5th edn, McGraw-Hill, Dubuque, Canada.

Savery, L. and Lucks, J. A. (2000) No productivity boost from downsizing, *HR News*, April, 1–2.

Saxon, R. (2005) *Be Valuable: A Guide to Creating Value in the Built Environment*, Constructing Excellence, London, UK.

Schaltegger, S. and Wagner, M. (2011) Sustainable entrepreneurship and sustainability innovation: categories and interactions, *Business Strategy and the Environment*, 20(1), 222–237.

Schon, D. (1983) Diffusion of innovation, in M. Lockett and R. Spear (eds) *Organizations and Systems*, The Open University Press, Stratford, England, chapter 19, pp. 192–202.

Schumpeter, J. A. (1936) *The Theory of Economic Development*, Trans Redvers Opie, Harvard University Press, Boston, Massachusetts, USA.

Schumpeter, J. A. (1947) *Capitalism, Socialism and Democracy*, 2nd edn, NY, Harper and Brothers.

Sebestye'n, G. (1998) *Construction Craft to Industry*, Spon, London.

Seneghi, F. and Loosemore, M. (2012): Employer-of-choice characteristics in the construction industry, *Construction Management and Economics*, 30(11), 941–950.

Senge, P. (1990) *The Fifth Discipline: The Art and Practice of the Learning Organization*, Doubleday, New York, USA.

Sexton, M. and Barrett, P. (2003) Appropriate innovation in small construction firms, *Construction Management and Economics*, 21(6), 623–633.

Sexton, M. and Barrett, P. (2005) Performance based building and innovation: balancing client and industry needs, *Building Research and Information*, 33(2), 142–148.

Sexton, M., Abbot, C. and Shu Ling, L. (2008) Challenging the illusion of the all powerful clients role in driving innovation, in P. Brandon and L. Shu-Ling (2008) *Clients Driving Innovation*, Wiley-Blackwell, Oxford, pp. 43–49.

Seymour, H. (1987) *The Multinational Construction Industry*, Croon Helm, London.

SFC (2002), *Rethinking Construction Accelerating Change,* Strategic Forum for Construction, Construction Best Practice Programme, London.

Sharad, M. and Swami, S. (2004) What factors influence pioneering advantage of companies?, *Vikalpa: The Journal for Decision Makers,* 29(3), 15–33.

Sharpe, W. (2004) Talking points: Managing stakeholder relations in PPP projects, *Public Infrastructure Bulletin*, March, 8–15.

Sheaffer, Z., Richardson, B. and Rosenblatt, Z. (1998). Early-warning-signals management: a lesson from the Barings crisis, *Journal of Contingencies and Crisis Management*, 6(1), 1–23.

Simon, H. A. (1956) Rational choice and the structure of the environment, *Psychological Review*, 63, 129–138.

Sinek, S. (2009) *Start With Why: How Great Leaders Inspire Action*, Penguin, New York.

Slaughter, S. (1983) Builders as a source of construction innovation, *ASCE Journal of Construction Engineering and Management*, 119(3), 532–549.

Slovic, P., Fischhoff, B. and Lichtenstein, S. (1981) *Perceived Risk: Psychological Factors and Social Implications*, Proceedings of the Royal Society, London, A 376, 17–34.

SMH (2010) The big country takes a lean turn, *Sydney Morning Herald*, 10–11 April, 4–5.

SMH (2012) Engineer who designed British space shuttle was ahead of his time, *Sydney Morning Herald*, 6 November, 18–19.

Smith, J. (2006) PFI: perfidious financial idiocy, *British Medical Journal*, 2(3), 319–321.

Smith, N. J. (1999) *Managing Risk in Construction Projects*, Blackwell Science, Oxford, UK.

Smith, P. (2011) Convergence is the future, *The Australian Financial Review*, special 60th birthday edition, 16 August.

Spence, M. (1981) The learning curve and competition, *Bell Journal of Economics*, 20(1), 49–70.

Spence, W. R. (1994) *Innovation*, Chapman and Hall, London, UK.

Steffan, M. (2012) 'Slaves' that are reflections of ourselves, *The Sydney Morning Herald*, 19 November, 7–8.

Steve, J. and Thomas, B. (2008) *Organizational Psychology: A Scientist-Practitioner Approach,* 2nd edn, John Wiley & Sons, Inc, Hoboken, New Jersey, pp. 341–365.

Stewart, M. (2009) *The Management Myth*, W. W. Norton and Co, Inc, New York.

Stickland, F. (1998) *The Dynamics of Change*, Routledge, London.

Sundbo, J. (1996) The balancing of empowerment: a strategic resource based model of organizational innovation activities in services and low tech firms, *Technovation*, 16(8), 397–409.

Sutton, R. I. (2001) The weird rules of creativity, *Harvard Business Review*, 79, 12–14.

Swann, W. B., Polzer J. T., Seyle D. C., Ko. S. J. (2004) Finding value in diversity: verification of personal and social self-views in diverse groups, *Academy of Management Review,* 29(1) 9–27.

Swedberg, R. (2006) Social entrepreneurship: the view of the young Schumpeter, in C. Steyaert and D. Hjorth (eds) *Entrepreneurship as Social Change*, Edward Elgar Press, Cheltenham, pp. 21–34.

Szilagyi, A. and Wallace, M. J. (1987) *Organisational Behaviour and Performance*, 4th edn, Scott, Forestman and Company, Illinois.

Taleb, N. N. (2001) *Fooled by Randomness: The Hidden Role of Chance in Life and in the Markets*, Random House Publishing Group, New York.

Taleb, N. N. (2007) *The Black Swan: The Impact of the Highly Improbable*, Penguin Books Limited, New York.

Taleb, N. N. and Blyth, M. (2011) Caught on the blindside, *Australian Financial Review*, 20 May, 10–11.

Tannant, S. and Fernie, S. (2011) Extended construction supply chain management: relationships, rewards and risks, in J. W. F. Wamelink, R. P. Geraedts and L. Volker (eds) *MISBE2011 – Proceedings of the International Conference on Management and Innovation for a Sustainable Built Environment*, Delft University of Technology, Amsterdam, The Netherlands, pp. 67–77.

Tarde, G. (1890) *Les lois de l'imitation*, translated by Elsie Clews Parsons in 1903 and published as *The Laws of Imitation*, H. Holt and Company, New York.

Tatum, C. B. (1987) Process of innovation in construction firms, *ASCE Journal of Construction Engineering and Management,* 113(4), 648–663.

Taylor, F. W. (1911) *Principles of Scientific Management,* Harper and Row, New York.

Taylor, J. (2005) *Unweaving the Rainbow: Research, Innovation and Risk in a Creative Economy*, AHRC Discussion Paper, Arts and Humanities Research Council (AHRC), London.

Teece, D. J., Pisano, G. and Shuen, A. (1997) Dynamic capabilities and strategic management, *Strategic Management Journal*, 18(3), 509–533.

Teo, M. M. M. (2008) *An Investigation of Community-Based Protest Movement Continuity Against Construction Projects*, unpublished PhD thesis, Sydney: FBE, UNSW.

The Economist (2013) Innovation pessimism: has the ideas machine broken down?, 406(8818), 12–18 January, 19–23.

Tidd, J. (2006) *A Review of Innovation Models*, Imperial College London, discussion paper 1/1.

Tjan, A. K., Harrington, R. J. and Hsieh, T. (2012) *Heart, Smarts, Guts, and Luck: What It Takes to Be an Entrepreneur and Build a Great Business*, Harvard Business Review Press, Boston, Massachusetts, USA.

Toffler, A. (1970) *Future Shock – A Study of Mass Bewilderment in the Face of Accelerating Change*, The Bodley Head, London.

Toole, M. (1998) Uncertainty and home builders' adoption of technological innovations, *ASCE Journal of Construction Engineering and Management*, 124(4), 323–332.

Townes, C. (1999) *How the Laser Happened: Adventures of a Scientist*, Oxford University Press, Oxford, UK.

Tversky, A. and Kahneman, D. (1974) Judgment under uncertainty: heuristics and biases, *Science*, 185, 1124–1131.

Tversky, A. and Koehler, D. J. (1994) Support theory: a non-extensional representation of subjective probability, *Psychological Review*, 101, 547–567.

Tzortzopoulos, P., Kagioglou, M. and Treadaway, K. (2008) A proposed taxonomy for construction clients, in P. Brandon and L. Shu-Ling (eds) *Clients Driving Innovation*, Wiley-Blackwell, Oxford, UK, pp. 58–67.

UKWON (2007) *High Performance Working*, UK Work Organisation Network (UKWON), London.

Venkatraman, N. and Ramanujam, V (1986) Measurement of business performance in strategy research: a comparison of approaches, *Academy of Management Review,* 11(4), 801–815.

Viereck, S. G. (1929) What life means to Einstein: an interview, *The Saturday Evening Post,* 26 October.

Wade, M. (2012) Get used to it: sending jobs overseas is the way of the future, *The Sydney Morning Herald*, 10 October, 1–2.

Walker, D. H. T. and Lloyd-Walker, B. (2011) *Profiling Professional Excellence in Alliance Management*, Alliancing Association of Australia, Melbourne, Australia.

Walker, D. H. T. and Rowlinson, S. (2008) *Procurement Systems: A Project Management Perspective*, Taylor and Francis, London, UK.

Wallach, M. A. (1985) Creativity testing and giftedness, in F. D. Horowitz and M. O'Brian (eds) *The Gifted and the Talented: Developmental Perspectives*, American Psychological Association, Washington, DC, USA.

Wang, C. L. and Ahmed, P. K. (2003) Organizational learning: a critical review, *The Learning Organization*, 10(1), 8–17.

Watson, T. J. (1994) *In Search of Management – Culture Chaos and Control in Managerial Work*, Thomson Business Press, London.

WBCSD (2012) *Vision 2050: New Agenda for Business*, World Business Council for Sustainable Development, Geneva, Switzerland.

Weber, E. U. and Hilton, D. J. (1990) Contextual effects in the interpretation of probability words, *Journal of Experimental Psychology*, 16, 781–789.

WEF (2013) *Global Risk Report 2013*, World Economic Forum, Davos, Switzerland.

Weick, K. E. (1976) Educational organizations as loosely coupled systems, *Administrative Science Quarterly*, 21(1), 1–19.

Weinstein, N. D. (1989) Optimism bias about personal risks, *Science*, 246, 1232–1233.

Wenda, B., Croissant, J. and Restivo, S. (2005). *Science, Technology, and Society: A Sociological Approach*, Wiley-Blackwell, Oxford, UK.

Werbac, A. (2009) *Strategy for Sustainability: A Business Manifesto*, Harvard Business Press, Boston, Massachusetts, USA.

White, S. (2012) Ageing boomers leave industry facing bust, *The Sydney Morning Herald*, 5–6 May, 19–20.

Widen, K. (2006) *Innovation Diffusion in the Construction Sector*, doctoral dissertation, October, University of Lund, Sweden.

Widen, K., Atkin, B. and Hommen, L. (2008) Setting the game plan – the role of clients in construction innovation and diffusion, in P. Brandon and L. Shu-Ling, *Clients Driving Innovation*, Wiley-Blackwell, Oxford, pp. 56–79.

Wilder, D. (1977) Perceptions of group size, size of opposition and social influence, *Journal of Experimental Social Psychology*, 13, 253–268.

Winch, G. (1998) Zephyrs of creative destruction: understanding the management of innovation in construction, *Building Research and Information,* 26, 268–279.

Winch, G. M. (2008) Revaluing construction: implications for the construction process, in P. Brandon and L. Shu-Ling, *Clients Driving Innovation*, Wiley-Blackwell, Oxford, pp. 16–25.

Wolstenholme, A. (2009) *Never Waste a Good Crisis*, Constructing Excellence, London.

Wood, Sir Kenneth (1975) *The Public Client and the Construction Industries*, HMSO, London.

Woodward, J. (1965) *Industrial Organization: Theory and Practice*, Oxford University Press, Oxford, UK.

Woudhuysen, J. and Abbey, I. (2004) *Why is the Construction Industry So Backward?*, Wiley, Chichester, UK.

Index